# THE AFRICAN INHE...

Africa is a continent gripped by civil wars, refugees and widespread famine. A significant number of the causes of these crises are deep-rooted, many extending back to the continent's colonial past when Africa was partitioned between the European powers. This European partition was to survive independence as the former colonies became separate sovereign states.

*The African Inheritance* discusses pre-colonial Africa, the increasing European interest in the continent, the European partition and the subsequent colonial rule and decolonization, and examines the consequences of its colonial inheritance: the large number of very small and weak states, the geographically marginal capital cities, contentious international boundaries, dependent land-locked states, destructive secessionist movements, irredentism and African imperialism. Africa has attempted to tackle these problems through abortive political union of states, economic groupings and reorientation of infrastructural development away from a colonially-based system. Yet individual development is inhibited by the colonially imposed strait-jacket of political geography.

In developing the theme of the colonial inheritance of Africa, *The African Inheritance* is essential reading for a better understanding of the confusing current problems of the continent. To look for immediate causes alone can be misleading. *The African Inheritance* directs the reader to a broader and deeper understanding of the contemporary map of the continent.

**Ieuan Ll. Griffiths** is Reader in Geography in the School of African and Asian Studies at the University of Sussex.

# THE AFRICAN INHERITANCE

*Ieuan Ll. Griffiths*

London and New York

First published 1995
by Routledge
11 New Fetter Lane, London EC4P 4EE

Simultaneously published in the USA and Canada
by Routledge
29 West 35th Street, New York, NY 10001

© 1995 Ieuan Ll. Griffiths

Typeset in Garamond by
Florencetype Limited, Stoodleigh, Devon

Printed and bound in Great Britain by
Biddles Ltd, Guildford and King's Lynn

*British Library Cataloguing in Publication Data*
A catalogue record for this book is available from the British Library

*Library of Congress Cataloguing in Publication Data*
A catalogue record for this book has been requested

ISBN 0–415–01091–8
0–415–01092–6 (pbk)

*To Margaret*

# CONTENTS

# MAPS

# ACKNOWLEDGEMENTS

This book takes my interest in African political geography a little further. As ever I am indebted to my colleagues and students at Sussex University for their varied contributions over the years, some made consciously with considerable effort, some made unwittingly, but nevertheless valuably. My special thanks again to Sue Rowland who has unfailingly transformed notes, sketches and vaguely expressed ideas into cartographic masterpieces. To the international referees my grateful thanks for their constructive criticism, not always appreciated at the time. To Tristan Palmer and his team at Routledge my thanks for their patient but unstinting support. Responsibility for errors of fact or judgement is mine. I hope that there are not too many.

# 1

# INTRODUCTION

For about a century, perhaps since 1885 when it was partitioned, Africa has been ruefully nursing the wounds inflicted on it by its colonial past. Remnants of this unenviable colonial heritage intermittently erupt into discordant social, political and even economic upheavals which, some may say, are better forgotten than remembered. But this 'heritage' is difficult, if not impossible to forget; aspects of it continue, like apparitions, to rear their heads, and haunt the entire continent in various jarring and sterile manifestations: how do you forget unhealed wounds?

(Ajibola 1994: 2)

From Cape to Cairo, Mozambique to Morocco, Somalia to Senegal, the continent of Africa is beset with life-threatening, large-scale problems. Famine and starvation, civil war and boundary bickering, crippling debt and crumbling infrastructure, plummeting economic performance and soaring population growth, burgeoning human disease and a devastated physical environment are features of everyday life in almost all parts of Africa. All is not well in the states of Africa. That is clearly understood, but there is far less understanding about the causes of the current ills of Africa and very few suggestions as to how those ills might be overcome in the long term.

All too easily the terrible effects are readily attributed to immediate causes. The refugees, the wars and the political instability are often the result of self-inflicted wounds of political graft and corruption, flawed democracies and power-hungry military dictatorships. Obviously, say the pundits, the ills of Africa are the result of a succession of shots in the feet and in that climate is it at all surprising that the shot wounds fester and gangrene threatens to overwhelm the whole body? But 'the unhealed wounds' of Africa's colonial past must also be remembered. The immediate causes of African misery must be put in the context of basic structural defects, both economic and political, deriving from the comparatively recent and short-lived colonial period when almost the whole of Africa was divided

1

between European powers. This context is by no means the sole cause of Africa's present plight, but the colonial inheritance is crucially important and not easily disowned.

The colonial experience is critical for modern Africa because of its strong impact on the present. Although clearly not the only influence on current affairs in modern Africa, and one that should not be interpreted deterministically, the imprint of the colonial inheritance is still very clear. Few radical changes have been made in the post-colonial period, to a large extent because of the strength, pervasiveness and durability of the colonial inheritance.

The significance of the colonial experience to Africa is out of all proportion to its duration which was generally surprisingly short. For most of Africa direct colonial rule was encapsulated within the seventy-five years 1885 to 1960. The range is very much wider. Ceuta, a small Spanish enclave on the north Moroccan coast, has been European-ruled continuously from 1415 to the present day. On a larger scale, parts of South Africa experienced colonial rule for over three hundred years, the extreme western Cape from 1652 to 1994. In contrast, Ethiopia, apart from Eritrea and the Ogaden, suffered only five years of European rule, 1936–41.

The key to the indelible impact of colonialism on Africa was the division of the continent into colonial territories. In the period 1885–1914 the European powers partitioned Africa between themselves. The partition was imposed on the continent with little regard to the distribution of peoples (ethno-linguistic or culture groups) or pre-colonial political units. Thereafter slight adjustments were made to that division of the continent by agreement between the European powers according to their own interests.

Crucially the European partition survived African independence almost intact. Independence movements were based on individual colonies and at independence the colonies became sovereign states. Pan-Africanism was an ideal nurtured essentially outside Africa, by exiles and by Afro-Americans in the widest sense. From America or Europe it was easy for an individual to identify with Africa as a whole, but on return to Africa that same individual inevitably became a Nigerian or would-be Ghanaian because the independence struggle was structured on individual colonial territories. Communication between territories, let alone co-ordination of independence movements, was difficult. This was particularly true of the British colonies in West Africa, where the pace for independence was set. They were isolated from each other as enclaves between French-held territory. Attempts to break this mould at the time of independence failed. Convinced Pan-Africanists such as Kwame Nkrumah tried but faced the basic contradiction that once independence was achieved, for say Ghana, it was more difficult to create practical and lasting union with another independent ex-colony, as was experienced in Ghana's attempted union with Guinea and Mali. Yet without independence there was no political freedom or power with which to aim for the Pan-Africanist goal anyway.

Independence on the basis of groups of colonies was more of possibility for French Africa where most territories were part of two large colonial federations, *Afrique Occidentale Française* (AOF) and *Afrique Equatoriale Française* (AEF). These had been formed early in the colonial period to satisfy the peculiarly French taste for administrative centralization and hierarchy. But the individual colonies were retained as a second tier of administration and although independence was almost collective in 1960 it was the individual colonies that became the new sovereign states. Britain created three groups of colonies, only one of which survived independence as a unit. The Union of South Africa, comprising four colonies, was created in 1910, in fulfilment of a British imperial dream of at least forty years standing, as a white minority-ruled state. The Federation of Rhodesia and Nyasaland was also white settler-dominated, formed against the wishes of a majority in the two northern territories who took the earliest opportunity to break away as separate majority-ruled independent states. In East Africa Britain created a series of common services jointly for the colonial territories of Kenya, Tanganyika and Uganda, but the arrangement fell well short of federation. Attaining independence at different dates, the territories went their separate ways. Later they came together briefly in a formal economic community. Was the European balkanization of Africa a deliberate, or force of circumstance, 'divide-and-rule' blueprint for neo-colonialism?

Whether the result of deliberate policy or not the colonially-imposed partition is a political strait-jacket for modern Africa. It divides, constrains development, encourages neo-colonialism and helps perpetuate Africa's economic and political weakness in global affairs. That weakness has many other causes but until the strictures of Africa's colonial inheritance are understood and relaxed there is no prospect of truly independent African development.

The partition of Africa allows insights into the present. The process of decolonization in Africa is not quite complete. In South Africa it was achieved only in 1994 with majority rule. The number of African people now under European rule is tiny but the political map of Africa is still not free from imperial colours.

The colonial inheritance contained two basic concepts alien to Africa: 'nation-states' and 'boundary lines'. In Europe nation-states and boundaries evolved over centuries with innumerable adjustments. In Africa the lines defining colonies were imposed mainly by means of bilateral agreements between colonial powers within a short period. Subsequent changes were mainly dictated by events outside Africa, rather than local conditions. The colonies created by this mesh of boundary lines were transformed at independence into would-be nation-states after a short colonial experience during which the sole unifying force within the colonial boundary was usually only the colonial adminstration itself, a point best illustrated by

3

forty-five (out of fifty-two) independent African states having imperial French, English, Portuguese or Spanish as their official language.

The European colonies were created by imposing a mesh of boundary lines on Africa with an almost total disregard for local conditions. Diverse culture groups were lumped together within a single colonial territory whilst at the same time individual culture groups were divided between colonies with indifference. Definite boundary lines replaced the more traditional borderlands.

Many of the states of modern Africa are too small, in terms of area, population and resources, for any real prospect of independent development. The vast area and ethno-linguistic diversity of other states has worked against cohesion and they have suffered political instability, almost as an in-built factor.

During European rule about 50,000 miles of colonial frontiers were imposed on the African continent. They are today international boundaries separating independent sovereign states. These colonially-inherited lines now define the states of modern Africa. They determine not only the size but also the shape of those states. The mesh of boundary lines is the fundamental basis for African development. The size and shape of states strongly influences their economic and political viability. This colonially-created mesh was endorsed by formal agreement of all but two (Morocco and Somalia) of the independent African states themselves, meeting as the Organization of African Unity (OAU) in Cairo in July 1964. All African states achieving independence since then have signed the agreement to respect their former colonial boundaries.

Along the great length of international boundaries in Africa there are many places where disputes have arisen in the post-independence era. Most of these disputes have had their origin in ambiguities, uncertainties and errors of judgement created and made by the colonial authorities. Most disputes have not been resolved with any finality, largely because of the 1964 OAU declaration. Some, such as the long-running dispute between Chad and Libya, where a change in the colonial boundary agreed between France and Italy was never formally ratified, have been resolved (in 1994) not by African arbitration, because no such mechanism exists, but by the International Court of Justice at The Hague. Disputes between sovereign African states could, perhaps should, be resolved by Africans, under the aegis of a revitalized OAU, if necessary by drawing new lines on the political map, or even by erasing some of the old ones.

Much economic and political activity in Africa is concentrated in the colonially-founded marginally-located capital city ports. These cities, placed as near to the metropolitan imperial power as it was possible to locate a colonial administration within the relevant territory in times of sea travel, are centres of chronic congestion, of which Lagos is probably the prime example. Combined with the extreme marginal locations of some of these capitals, such congestion stifles development. Some new, more centrally relocated capital cities have already been built in modern Africa. They are as much symbols of independence and national unity as vehicles for accelerating and spreading economic development. As solutions to colonially-created problems they are always very costly and heavily risk-laden.

An important part of Africa's colonial inheritance lies in its fifteen land-locked states, more than in any other continent, even after the break-up of the former Soviet Union. What were perfectly acceptable to the imperial powers as remote, thinly and sometimes indirectly ruled colonial protectorates or military territories at the end of long lines of imperial communication have now been transformed into the very different reality of sovereign states. They fear dependency on their seaboard neighbours and for the security of their trade life-lines which give access to the sea. Already in post-independence Africa some alternative routes have been constructed from land-locked states to the sea-coast, but again these are costly solutions. They also represent problems not clearly perceived as worthy of costly solution by former imperial powers now cast in the role of money-lenders and aid-donors. Projects, such as the railway from Zambia to Dar es Salaam, viewed as political necessities from the peculiarly continental viewpoint of an African land-locked state, have been turned down for assistance by Western countries seeing the situation quite differently from their own outsider perspective with a priority of economic cost-effectiveness. Many of the land-locked states in Africa have suffered closed borders and have had their particular vulnerability brought home to them by neighbouring sea-board states insensitive to rights of access to the sea.

Secessionist movements in independent Africa have been met by nothing more constructive than incantations of territorial inviolability by every colony-become-independent-state and by generally successful claims that they are exclusively internal affairs which brook no external interference. As a result there are many civil wars raging in Africa, contributing greatly to the continuing poverty of the continent. Some of these wars have been waged more or less continuously for more than thirty years and owe their origin to decisions taken about Africa by outside powers. In the case of Eritrea, the perceived strategic interests of the outside power, the United States, in supporting Ethiopia was paramount rather than the right of the people of the former Italian colony of Eritrea to self-determination, for all the fine words of the United Nations' Charter. The two secessionist wars which arguably caused most suffering to ordinary Africans came early in the post-independence era, in the Congo (Zaire) in the early 1960s and in Nigeria at the end of that decade. Both had appreciable non-African involvement which was blatantly neo-colonial as both wars were closely linked with the availability of rich mineral resources, respectively copper and oil, in the potentially secessionist areas of Katanga and Biafra.

5

Irredentism, the claim to unite territory on the grounds of a common language, is a rare phenomenon in Africa because pre-colonial ethnolinguistic groupings were generally far less extensive than the states of modern Africa. But the Somali irredentist cause was strong long before independence in 1960 united the two former colonies of British and Italian Somaliland but still left about one-third of all Somali speakers under the rule of three neighbouring territories, Djibouti, Ethiopia and Kenya. Pre-independence attempts to bring all Somalis together in a single state failed when that aim was seen to run counter to the interests of the emerging world super-powers of the United States and the Soviet Union. Somalia was one of only two African states not to sign the 1964 OAU agreement that all colonial boundaries in Africa should be respected because it could not accept that lines which divide Somali people should be given any status. Swaziland represents the extreme case of irredentism where the ethnically homogeneous population of that independent state is actually outnumbered by members of the same ethno-linguistic group resident across the border in the neighbouring Republic of South Africa. Attempts to incorporate these people (without consultation) into Swaziland by greatly increasing the land area of the tiny kingdom in the proposed 'land deal' with South Africa failed after 1984 when the deal no longer fitted in with the strategic machinations of the apartheid state.

Africa then has not only suffered sorely from having imposed upon it the priorities of European colonialism but also the more recent external influence of neo-colonialism. Such interference has had deleterious effects on the political geography of modern Africa not least, for example, by encouraging the emergence, or rather re-emergence, of home-bred African imperialism. Ethiopia and Morocco, their governments strongly supported by the Soviet Union and the United States respectively, aimed to subjugate the people and territories of Eritrea and Western Sahara respectively. For good measure, in the case of Western Sahara, exploitation of mineral resources (phosphates) is a significant element in a Moroccan imperialistic venture which has little to distinguish it from the European variety from which Africa suffered for so long.

In attempting to overcome the influences of colonialism and neo-colonialism Africa got off to a poor start. Early efforts failed to create a radical alternative to individual colonies becoming, in the main part, rather weak independent states, peculiarly vulnerable to the forces of neo-colonialism. Conservative forces, rooted in established regimes in the continent with some external backing, prevailed against ideas of pan-Africanism which were largely born and nurtured outside Africa. The words of Kwame Nkrumah, the first leader of independent Ghana, that 'Africa must unite' went unheeded and the radical solution, simplified as 'political union before economic union', was put aside for the conservative solution, simplified as 'economic union before political union'. There was to be no

'United States of Africa', instead in 1963 at Addis Ababa the Organization of African Unity (OAU) was set up, an international club with membership restricted to the majority-ruled independent states of Africa. The undistinguished record of the OAU with hindsight confirms that movement for what it was interpreted to be at the time of its creation, a defeat for the forward-looking radicalism of Nkrumah and other like-minded anti-colonialists. In Somalia (1960) the strength of irredentist feeling led Britain and Italy to bring their separate colonies together in a single state. Other attempts at political union between states followed independence but only the union between Tanganyika and Zanzibar (1964) has survived. Otherwise individual colonies were simply transformed into individual sovereign states and they have stayed that way. For the most part African states are minor political entities on the world stage, with a large fringe of tiny statelets with minuscule economies and no political clout.

Small states with tiny economies are ineffective units for independent economic development. This was recognized by the imperial powers when they federated colonies in Africa but then was all too conveniently forgotten as the colonies achieved independence. To bind established independent sovereign states into close economic co-operation with each other is an extremely difficult task as the on-going experience of the European Union shows. When the states attempting to come together in economic co-operation are newly independent and need to stress their identity in order to achieve a basic unity; are poor, with economic infrastructures both patchy, low-level and essentially internalized; and also suffer from enormous internal and external disparities in levels of development, then the task is very much greater, almost to the point of being impossible.

In East Africa the British left a legacy of well-established common services. Although the British did not follow the logic of this co-operation through to a full union of its four colonies (Kenya, Tanganyika, Uganda and Zanzibar), and did not even take up Julius Nyerere's suggestion of a common independence date for all the colonies, the East African Community (EAC) was established within four years of independence (in 1967). But it was very short-lived. In 1977 the EAC collapsed, a victim of economic and ideological disparities, political instability and personality clashes. Its demise, long forecasted and first signalled by the closure of the frontier between Kenya and Tanzania, was followed in 1979 by war between the Uganda of Idi Amin and Tanzania.

In West Africa the Economic Community of West African States (ECOWAS) has made heavy weather of economic co-operation. Nigeria dominates ECOWAS and as the economic core of the region attracts hundreds of thousands of migrant workers from a periphery of poorer surrounding states. Yet on two occasions the Nigerian government has expelled non-Nigerians *en masse* in an effort to reduce local unemployment when the economic going got tough after the slump in oil prices in the

1980s, so undermining confidence in the organization and the ability of the member states to co-operate in development. In the 1990s a deliberate attempt has been made to give ECOWAS a political dimension by committing a military force to Liberia to hold the ring in the devastating civil war and in particular to help prevent the conflict spilling over into neighbouring states. The intervention, largely manned and financed by Nigeria, has not been a complete success, though final judgement must be reserved.

South Africa traditionally dominated the sub-continental region of southern Africa throughout the colonial period. Its mining-led modern economy has attracted migrant workers from most parts of the region, mainly on short-term contracts. Denied the opportunity to move permanently by restrictions imposed by colonial authorities and more emphatically by apartheid, these workers became part of the region's dual economy, belonging to the periphery but admitted to the core only when, and for as long as, their labour was needed. The system is one of the more pernicious aspects of the colonial inheritance in Africa.

Set up in part to break the colonial system and specifically to encourage development in the peripheral states *independent* of South Africa is the Southern African Development Community (SADC), formerly the Southern African Development Co-ordination Conference (SADCC). It has so far proved to be an effective aid umbrella, especially for transport development, but little else. SADC now faces a major challenge over South Africa, because majority rule there calls for a complete rethink of the purposes of SADC. It may prove more difficult for the land-locked states of southern Africa to develop independently of a friendly South Africa than it was of a South Africa which was blatantly hostile. The dominant position of South Africa within the sub-continent, with 20 per cent of the land area, 40 per cent of the population and 80 per cent of the wealth in terms of Gross National Product (GNP) is certainly going to be difficult to handle. The very existence of a post-apartheid South Africa is another aspect of colonial inheritance that will extend the range of economic disparity among SADC states and so render the practicalities of co-operation more difficult. The Preferential Trade Area (PTA) incorporates most of the states of southern and eastern Africa from Zimbabwe to Kenya. Its progress has also been slow and, as with SADC, the nettles of economic co-operation have not been grasped. In the light of the experience of the EAC a gradualist approach may be prudent, but forward movement in either SADC or PTA is barely, if at all, perceptible. The PTA also has to confront the implications of extending membership to a majority-ruled, but still economically dominant South Africa.

Colonially-developed lines of communication usually ran between inland resource locations, such as mines or plantations, and the ports. Transport networks were often contained within the frontiers of a single colony, and even within individual colonies lateral transport links were limited. The infrastructure necessary for the development of intra-African trade, economic co-operation and development co-ordination demands priority within modern Africa although the immediate returns are likely to be very small. Once installed, road and rail links can be put to many uses, not all of which would have been seen clearly at the time of construction. It is perhaps ironic that the largest multi-national transport network on the continent, that of southern Africa, was used in the 1980s by white South Africa in an attempt to assert political and economic hegemony over its neighbouring states. That network comprises railways originally built for colonial mineral exploitation, a rather different form of political domination than that exerted by white South Africa in the 1980s. It may be that in the post-apartheid era that same transport network can be put to more constructive use to the benefit of all the countries of the region. If Africa waits for intra-African trade to develop before making the infrastructural investment necessary to facilitate that trade the wait is likely to be long indeed. Yet projects such as the trans-Africa highway, aimed at improving transport links between West and East Africa, have been given very low priority. Africa has been consigned by the economically powerful industrialized countries to the lowly position in the world economic order of raw material provider. To contest this, through self-reliance in the form of intra-African trade and development, Africa must place intra-continental, international, infrastructural improvements high on the agenda.

The colonial experience was not all bad for Africa and the post-colonial experience has not been all good. Thirty years or more, a generation, into independence, Africa cannot blame all its troubles on its colonial past. Even though any attempted assertion of true independence is liable to be undermined by the forces of neo-colonialism, progress is also hampered by Africa's seemingly endemic political immaturity. Only one year (1989) has gone by in post-independence Africa without at least one *coup d'état* succeeding in toppling an established government. Whilst arguments rage as to the appropriateness for Africa of multi-party or single-party democracies, most countries in Africa are ruled by the military, small political elites or individual dictators. There have been too few initiatives to promote truly African interests and too many willing African accomplices of neo-colonialism, greedy for the spoils of political power and commercial graft. To make matters worse, progress has also been hindered by the mis-directions even of well-intentioned leaders who, often taking their cue from outside, have given priority to wrong-headed schemes. But then there is a learning process through which all have to go and African leaders have to be allowed to make their share of mistakes without a chorus of criticism from outsiders, especially when many of the worst offenders, Amin, Bokassa, Mobutu and Savimbi have been encouraged in their excesses by outsiders.

More than one hundred years after Bismarck's Berlin Conference of 1884–5 laid the ground rules for the division of Africa between the European powers the political geography of African development demands the closest of attention. The time is ripe to ask new questions of African leaders, to challenge some of the sacred cows of African independence whilst not forgetting that the framework of political geography within which Africans have had to shape their independence was imposed rudely and for a very different colonial purpose by Europeans. Africa has also laboured under the on-going constraints of neo-colonialism. Although much of Africa has been nominally independent since the 1960s the continent has continued to be drained of its resources and deliberate under-development is unabated. Ultimately the problems of Africa can be solved only by Africans, including white (Afrikaner) Africans. Outsiders, whether from Europe, the United States of America or elsewhere can usefully promote discussion of Africa's problems, can put forward suggestions and make positive contributions when asked. Although, indeed because, outsiders collectively did and continue to do so much to create problems for Africa, they should now stand aside from decision-making in the continent whilst being ready to assist in making good some of the damage inflicted on the continent of Africa over the colonial and neo-colonial periods.

# 2

# PRE-COLONIAL AFRICA

The period of European rule in Africa by the British, French, Germans, Italians, Portuguese and Spanish, needs to be put into perspective. It was historically short and was only one of a large number of outside influences experienced by Africa. The impact of European rule on Africa is great because: it covered virtually all the continent; it was accompanied by a technological revolution leaving a legacy from railways and aircraft to factories and hydro-electric dams; it introduced capitalism and gave Africa a lowly place in a world economic order; and it is recent. But Africa had a very long and complex history of its own before the relatively short period of European colonization. In addition parts of Africa, particularly the north and north-east, were subjected to the rule of several different empires based outside Africa: Persian, Greek, Roman and Ottoman. In addition Islam, the Moors and the Swahili trading empire have been, and are to the present day, other strong outside influences brought to bear on Africa.

Africa has sound claims to being regarded as the cradle of mankind. It is the only continent on which hominid remains of the type *Australopithecus* have so far been found. The first discovery, in a limestone quarry at Taung in the northern Cape Province of South Africa, was followed up by discoveries of similar hominid remains in the Transvaal in the late 1930s, and since 1959 in the Rift Valley of East Africa where the Olduvai Gorge is the best-known site. The geographical conundrum is that such remains have not yet been found anywhere else on earth except Africa. It might be that such remains, which plug the gap between the universally-found *Ramapithecus* of 12 million years ago and *Homo erectus* of about 1 million years ago, are awaiting discovery somewhere outside Africa, but so far only tantalizing artefacts have been brought to light.

The lower Nile valley was the site of one of the great civilizations of the ancient world. Here, in favoured natural conditions, with the great river providing flood irrigation combined with rich nutrient replenishment in the form of silt, a large and regular food surplus could be grown by only a small proportion of the total population, so permitting the growth of a complex society and polity. The achievements of Ancient Egypt were remarkable, for example, the obvious architectural pinnacles of the pyramids, the Sphinx and the temples. Even a short visit to modern Cairo, Giza and the Museum of Antiquities makes one realize the extent and depth of the civilization that created these monuments – achievements in writing, even paper manufacture, medicine and science, mathematics and engineering and perhaps above all in the arts. Not only was Egypt one of the earliest and richest civilizations, it was also one of the most durable, lasting almost 2,500 years from before 3000 BC to the Assyrian conquest of 665 BC, a lifespan against which the European colonization of Africa, encapsulated well within a single century, needs to be considered. The civilization of Egypt was centred on the Nile valley below Aswan and the great delta. From this core the Egyptian empire at its height extended north to Syria, west to Cyrenaica and south to Nubia, commanding the eastern Mediterranean and the Red Sea at the great strategic cross-roads of both ancient and modern worlds. The Egyptians first conceived of building a canal to allow passage of ships from the Mediterranean to the Red Sea but abandoned the task, instead sending Phoenician sailors to circumnavigate Africa from east to west on their behalf, a stupendous feat which took over two years to accomplish.

The Assyrians were the first of a succession of Eurasian empires to conquer Egypt and other parts of north Africa. In 525 BC the Persians under Cambyses conquered Egypt and Cyrenaica. Four years later Darius completed the canal from the Gulf of Suez to the River Nile and thereby opened a direct sea route from Persia to the Mediterranean via the Red Sea.

The Phoenicians not only rounded Africa but regularly covered the length and breadth of the Mediterranean in their trading voyages. They ventured beyond the Pillars of Hercules (the Straits of Gibraltar) to set up small trading posts on the Atlantic shoreline. In North Africa they founded what became an independent trading empire based on Carthage, from which later Hannibal with his elephants was to lend a special African flavour to his daring trans-Alpine challenge to the might of Rome.

The Greeks also traded across the Mediterranean and established posts in North Africa. The most famous was the city of Alexandria, founded in 332 BC by, and named for, Alexander the Great, at the western end of the Nile delta coast. Alexandria occupies the classical location of a colonial port-capital city, at the point in the colony nearest the metropolitan country. Greek scientists, geographers and historians were active in Egypt. Ptolemy speculated about the source of the Nile and calculated the size of the earth from angles of the sun's declination at different times of the year at the Tropic of Cancer, whilst Herodotus recorded what had previously been oral history of the ancient empires. Greek settlements were established as far down the Red Sea coast as Adulis in modern Eritrea.

The Roman Empire eventually took in the whole of North Africa. Carthage, under Hannibal, was defeated at Zama in modern Tunisia in 202 BC, but it was not until 30 BC that Egypt fell to the legions of Julius

Caesar. North Africa was part of the Roman empire for almost five hundred years until the Empire itself fell, again a timespan against which to measure the European colonization of Africa in the late nineteenth and twentieth centuries. The East Roman Empire based on Constantinople then continued to hold sway over Egypt. Under this Christian empire Egypt and Nubia were converted to its religion.

In 632 AD the Prophet Mohammed died at Medina in present-day Saudi Arabia. Within ten years, inspired by the new religion of Islam, the Arab Caliphate had taken Egypt. Armed with their new religion the Arabs rapidly spread along the Mediterranean coast of North Africa, reaching what is now Morocco in about 705 AD. From there the Moors crossed into Spain to occupy the southern half of the Iberian peninsula before being driven out by a revitalized rival religious militancy in the form of small Christian kingdoms.

Islam spread down the Red Sea littoral, across the Horn of Africa in the eighth century, and southwards along the sea-trading route of the Swahili coast. Penetration of the interior from the Red Sea coast was held up by the Christian bastions of Nubia and Abyssinia, until the former was over-whelmed and the latter, protected by its rugged terrain, became an enclave, an island of Christianity in a sea of Islam. The caravan trading routes across the Sahara carried Islam to the savannah areas of West Africa by the eleventh century, spreading east–west along the Sudanic belt between the desert and the tropical forest area to the south. Islam converted and contributed to the development of the Sahelian states. Farther south, tropical Africa was not greatly affected by Islam, perhaps defended by climate, forest and disease. With just a few notable exceptions, such as the upper Nile valley where it made further progress in the nineteenth century, Islam reached its present limits by about 1500 AD. Unfortunately many European colonies in Africa, and the successor modern African states, in straddling the forest, savannah and desert belt, also straddle the Muslim/non-Muslim divide. By ignoring this great cultural divide the colonial powers created flawed political units. The division not only makes national unity more difficult to achieve in states such as Nigeria, Chad and Sudan but has actually been a contributory cause to civil war. Nowhere is this more evident than in the Sudan where the Muslim/non-Muslim divide is the basic cause of the on-going civil war which has bedevilled the Sudan for almost all of its existence as a modern African state since independence in 1956.

As the inspiration of Islam drove the Turks to take Constantinople (1453) and on to the gates of Vienna, so they established their Ottoman Empire throughout much of the Mediterranean and North Africa in succession to the Greeks and the Romans. Much the same area of Africa came under yet another form of alien, imperial rule.

In Africa south of the Sahara, large-scale long-distance population migra-tions were taking place. From a starting point in West Africa, Bantu-speaking

peoples moved east and south, reaching the area of the great lakes in East
Africa and then, following the fairly open savannah country around the
tropical rain-forest of the Congo basin into southern Africa where they came
into contact with the Khoi-khoi (Hottentots) and San (Bushmen). The
Bantu-speakers were essentially cattle herders, as were the Khoi-khoi, whilst
the San were hunters and gatherers. Pressured by the greater numbers and
superior technology of the Bantu-speakers, the Khoi-khoi and the San were
gradually pushed into the south-west corner of the southern African sub-
continent. The history of these groups was not recorded but from early
European mariners' written reports it is clear that the Bantu-speakers had
reached the north-eastern parts of the present-day South African coast by the
time of the early Portuguese voyages which again circumnavigated Africa,
this time from west to east, in the quest for the European sea route to India.

Along the Sahelian corridor between the Sahara desert and the rain-forest
areas of West Africa there arose a number of empires, the best-known
of which were Mali and Ghana, names adopted by the modern African
states, with some geographical licence, to celebrate the existence of a
political African past extending back over a millenium. The Sahelian states
were essentially trading empires situated at the junctions of north–south
and east–west long-distance trading routes. Their trade included the West
African products of gold, ivory, slaves and salt, the last from the Sahara
itself. Across the Sahara came products of the Mediterranean, whilst the
fertile lands of the inland delta of the Niger were well able to produce
enough surplus food to support large urban populations at trading cities
such as Timbuctoo and Gao. The wealth and sophistication of these empires
was considerable, impressing visiting travellers and states visited by their
rulers alike.

A form of Islamic fundamentalism flourished even in eleventh-century
north-west Africa, and the Almoravid sect, based in the area of present-
day Mauritania, overwhelmed much of the western Sahel. The fatal threat
to the Sahelian states also came from the north-west when, in 1591, the
Songhai empire based on Timbuctoo was defeated by the Moroccans whose
army had accomplished the remarkable feat of crossing the Sahara to attack
the fabled city. Morocco held fiefdom over the western Sahel, installing
a Moroccan governor and army of occupation. As the centuries passed
political links with Morocco became more tenuous, but the Moroccan
hegemony established in the sixteenth century was the chief basis of
Moroccan territorial claims over vast areas of north-west Africa since
modern Morocco's independence from France and Spain in 1956. The
Moroccan invasion contributed to the long-term decline in the prosperity
of the western Sahelian trading centres as did the establishment of European
trading-posts along the West African coast which diverted trade away from
the Sahel. Cities, such as Timbuctoo, no longer occupied the centre ground
but became peripheral to the main trading areas to north and south and

therefore declined to the relatively impoverished state they had reached by the time Europeans first ventured into the West African interior in the nineteenth century.

Farther east along the Sahel other states, centred on Hausa-land, Bornu and Darfur flourished on much the same sort of combination of trade routes and diverse regions capable of supporting large populations. The power of individual Sahelian states waxed and waned but they were fairly sophisticated political organizations with complex societies and economies, and they existed long before Europeans penetrated the interior of Africa.

Small but significant states emerged in the forest fringe areas near the West African coast. Here, unlike much of the rest of sub-Saharan Africa, there was an urban tradition and all that implies in terms of ability to produce a regular food surplus and social and political organization. It is claimed that the Yoruba, for example, lived in towns, as opposed to cities such as Timbuctoo, but nevertheless these were very much indigenous African creations demanding complex organizational skills. The Ashanti, of present-day Ghana, presided over a gold-rich economy, and included the crafting of exquisite gold-weights in their cultural achievements. Farther east at Benin, in present-day Nigeria, another sophisticated metal-working craft was practised in the production of bronze castings of such magnificence that many Europeans, on seeing them for the first time, could not accept that this was work indigenous to Africa and, not for the last time in Africa, attributed this work to outside agents (usually of European origin).

Between the great lakes of East Africa there arose a number of kingdoms among the Bantu-speaking peoples, mainly in the area of present-day Uganda. Chief among them was Buganda, but it lived in uneasy rivalry with its neighbours of Bunyoro, Busoga, Ankole and Toro. This was another extremely rich and fertile area, not this time one of the great river valleys but an area where the presence of the great lakes modified the local climate to produce abundant and regular rainfall and where tropical heat was modified by an elevation of about 3,000 feet above sea level. Perennial cropping was (and is) possible and produced regular food surpluses, so enabling the development of complex societies quite independent of external formative influences. The nature of these truly African societies was fully reported, but not always fully appreciated, in the nineteenth-century accounts of the first European explorers on the scene, such as James Hanning Speke.

All around the tropical rain-forest core of Africa from present-day Uganda to the northern parts of present-day Angola were Bantu-kingdoms of differing strengths and accomplishments. Variations were mainly due to local resources and the ability to produce regular food surpluses. These societies worked metal, mainly iron and copper, and cultivated the soil as well as rearing cattle. Their histories are not fully known as the telling has to relate to oral tradition because they laboured under the major disadvantage of not having a written language. They were also handicapped in having a very limited technology, in which the wheel was conspicuous by its absence. Although draught oxen were used it was to haul crude sledges which were severely limited in terms of transport speed, distance and load. The absence of the wheel had wider effects, so that although pottery was made throughout Africa south of the Sahara, it was not made on the wheel, which was unknown there. To this day African pots are coiled, though they have as perfect a symmetry and fineness as if they had been fashioned on the wheel. It was in the area of technology that African societies suffered when they came into contact and rivalry with Europeans. Nowhere more so than in the area of weaponry, where the technological gap became, in the late nineteenth-century, at least as great as that demonstrated between the two sides in the Gulf War of 1991. The superiority of machine guns, first the Gatling gun and then the Maxim gun, over the assegai was total, making so many of the 'colonial wars' of the late nineteenth century sickeningly one-sided. Nevertheless African societies were generally much better organized politically and socially than most Europeans who came in contact with them credited. They had also developed skills in mining, metal-working and even stone construction to such a high degree that many Europeans simply could not accept the obvious and looked for fanciful exotic explanations for the presence of such skills.

Based on very extensive gold mining with a history of many centuries, the empire of the Monomatapa in present-day Zimbabwe was one of the most interesting and enigmatic societies of pre-European Africa. It was an enigma because when Europeans first visited Zimbabwe in numbers during the latter half of the nineteenth century the Monomatapa no longer existed and there was little to distinguish Mashona society from that of other Bantu-speaking groups. Indeed it appeared to the European hunters, traders, missionaries, prospectors and the other forerunners of Rhodes' pioneer column which entered the country in 1890, that the Mashona owed allegiance to the Ndebele based at Bulawayo in the west of the country. Such apparent subservience did not encourage thoughts that the Mashona were the remains of the empire of the Monomatapa. Yet there was the evidence, literally thousands of old mine workings throughout the south and east of the country along auriferous and occasionally copper reefs and as many stone-built ruins, the most spectacular of which were the ruins of Great Zimbabwe from which the modern state takes its name. How to reconcile the archaeological evidence of the mines and ruins with the sociological evidence of a very ordinary African people?

The mine workings were generally shallow open-cast pits, usually between 30 feet (9m) and 50 feet (15m) deep, occasionally double that, also in trenches averaging between 150 feet (45m) and 600 feet (185m), but occasionally up to 4,500 feet (1,385m) in length. There were also inclined adits driven into the sides of the hills. This pointed to limitations imposed by a lack of technology to cope with problems of water drainage. But the number of ancient workings, in excess of 75,000 at a conservative estimate, and the amount of material removed, at another conservative estimate in excess of 43 million tons, was such that mining must have gone on for a very long time. The output from these mines must have been very considerable, again estimated conservatively as well in excess of 20 million ounces of fine gold.

The stone-works were first closely examined by outsiders in the late nineteenth century. First by the German geologist, Mauch, in 1871, and in 1891 by the American explorer and archaeologist Theodore Bent. They attracted close attention, being both fascinating as structures and difficult to explain. Their function was not clear yet their complex construction called for the very highest stone-working skills. They were built of carefully dressed stones, sometimes elaborately patterned, though the quality of the stone-work varied from place to place and even within the same building. At Great Zimbabwe, which was only the largest and most elaborate of very many such ruins, there was an outer stone wall, varying in height between 15 and 35 feet (4.5m to 10.5m) and about 16 feet (5m) thick at the base tapering to about 5 feet (1.5m) at the top. In plan this wall was roughly elliptical in shape and had an overall external length of about 830 feet (255m). For about one-third of this length there was an inner double wall, separated from the outer wall by a narrow passage. Inside the enclosure there were several structures, round tapering towers, a raised platform, stone stairways, inner walls and standing stones. The outer walls were pierced by three narrow entrances, one cut obliquely, and by one 8-inch (0.2m) square aperture 4 feet (1.2m) above the base of the 16-foot (5m) thick wall.

Bent set the pattern for explanation of the mysteries of Zimbabwe, filling his book with allusions to what he saw as similar structures in the Middle East. He also named the various structures, such as temple, altars and acropolis; he recorded various angles to the meridian and took careful measurements, sometimes expressed in exact cubits, and related the height and circumference of buildings to distances between buildings. He came to the conclusion that Great Zimbabwe was the work of the Phoenicians, leading to the explanation that Zimbabwe was indeed, as even John Milton had claimed in *Paradise Lost* (xi, 399–901), none other than the Biblical land of Ophir, the location of King Solomon's mines and the home of the Queen of Sheba (and Rider Haggard's *She*). Hall and Neal, writing ten years later, carried the Phoenician connection to the extreme of alluding to the *Book of Job* (xviii, 1–11) in reference to their mining. Where the ancient mine workings showed evidence of fairly recent timber-work they concluded that this 'must have been erected by the Portuguese in the seventeenth century' (Hall and Neal 1902: 55). The curious inability of non-Africans to accept that the mines and stone-works of Zimbabwe were the work of Africans continues, and as late as 1988 a book was published to reassert the Phoenician connection, supported by new hypotheses as to the function of various buildings in the Great Zimbabwe complex.

17

From Arab and Portuguese sources it seems evident that the gold mines of the Monomatapa produced vast quantities of gold at least up to the end of the fifteenth century. The down-turn in production seems to have coincided with the Portuguese take-over of the Arab trading posts on the Indian Ocean coast, the most important for gold shipments being Sofala. There is no mention of Phoenicians or any other foreigners then leaving Zimbabwe or any subsequent reference to a Phoenician or any other foreign element being present in the population of Mashonaland. Without doubt the gold of the Monomatapa was traded on the Indian Ocean coast and much of it did find its way to the Middle East. Indeed there could be a truthful basis to the myth of the land of Ophir but that would not necessitate the ancient miners and stone-workers being anything other than African.

The early nineteenth century was a period of nation-building among the Bantu-speaking people of present-day South Africa. In that country there were two branches of Bantu-speakers, the Nguni and the Sotho. The former occupied the land between the coast and concordant Great Escarpment, the latter the high veld area above the Great Escarpment. Among the Nguni people in the area of present-day Natal, pressure on grazing lands caused by a period of drought years and famine succeeding years of above-average rainfall seemed to be the main cause of conflict between various chiefdoms which resulted in far-reaching social and political change.

Alliances between chiefs resulted in the emergence of three kingdoms to contest domination of the area. The weakest group, the Ngwane under Sobhuza, removed themselves a little to the north where, by overcoming other smaller chiefdoms, they eventually succeeded in establishing the modern Swazi nation. At the epicentre there remained two powerful groups, the Mthethwa under Dingiswayo and the Ndwandwe under Zwide. In an ensuing battle Dingiswayo was killed, to be succeeded by Shaka, petty-chief of the Zulu sub-group. Shaka proved to be a much more effective opponent who not only revolutionized the military tactics of the time but also succeeded in changing the social structure of the Zulu people to facilitate the employment of the new military tactics.

Shaka's *impis* were organized in age-groups which were kept in continuous army service and housed together in special military *kraals*. They achieved greater military discipline and group loyalty under the system of continuous military service, which also assisted in welding together as a single fighting force peoples of many diverse sub-groups. These changes resulted in a much greater centralization of power in the hands of the king.

The new military tactics involved the use of a short stabbing spear and a large cow-hide shield. The disciplined *impis* raised a wall of shields to give protection against initial volleys of traditional long spears thrown by the enemy who, then weaponless, were engaged at close quarters by Zulu warriors armed with the short stabbing spears, who also employed the heavy

18

*Map 1* Pre-colonial Africa

shields as weapons. The Zulu also advanced in disciplined fashion in an ox-horn formation, a main body with two out-flung flanks. Once the main body at the centre engaged the enemy the horns continued to advance in a pincer movement to encircle the enemy. The key to the new tactics was rigid military discipline, and it proved to be of devastating effectiveness against all comers.

Shaka and his Zulu defeated the Ndwandwe under Zwide in a decisive pitched battle on the Mhlatuse river in 1818. Having established control of the epicentre Shaka proceeded to build up the Zulu nation from the many diverse sub-groups of people in the region. The task was more or less fulfilled when in 1828 Shaka was assassinated by his two brothers, one of whom, Dingane, then killed the other and himself assumed the Zulu throne. Within a very short period the Zulu had emerged from being a small sub-group within Dingiswayo's Mthethwa to become the dominant group among all the Nguni peoples. The Zulu nation gained in size and strength through its aggression and ability to absorb its conquered foe. It has proved durable, and has survived to this day as the largest single group of South Africans, much more numerous than and as distinctive as the Boers themselves, despite heavy defeats by the Boers in 1838 and the British in 1879.

Many chose not to be forcibly absorbed into the Zulu nation and took the one alternative available to them, to flee the epicentre. This process started long before Shaka's defeat of Zwide, but in total the *Mfecane*, a series of mass migrations of people accompanied by conflict, upheaval and the emergence, through absorption, of new 'nations', was to affect much of the eastern side of southern Africa, from the Eastern Cape Province of South Africa as far north as Lake Victoria and as far west as the upper Zambesi valley. The movements were not confined to the Nguni people because as they spilled over onto the high veld to seek refuge from the Zulu epicentre they started up chain reactions among Sotho peoples who displaced other groups as marauding bands devastated wide tracts of southern Africa.

Among nations forged at this time were the Swazi, who had early escaped the major conflict but remained perilously close to the Zulu epicentre across the Pongola river. On the high veld the modern Sotho nation was built up by Moshoeshoe. He united the survivors of many devastated groups under his brilliant political leadership and cleverly used the rugged terrain to good military effect against vengeful Zulus and encroaching Boers alike. About 1826 one of Shaka's *indunas*, Mzilikase was forced to flee Zululand with his people. Harrassed by Shaka's *impis*, Mzilikase and his group, who became known as the Ndebele, cut a great swathe through the high veld of the present-day Transvaal, absorbing many Sotho and Tswana communities as they went. They settled in the Marico basin of the western Transvaal until they were expelled by the trek-Boers in 1838. They again fled, this time to the north, eventually settling in Matabeleland in south-western Zimbabwe. From his royal *kraal* at Bulawayo Mzilikase established a hegemony over most of present-day Zimbabwe, including Mashonaland.

In addition to these nations who have survived subsequent European colonialism as major forces in southern Africa, smaller groups of people displaced during the *Mfecane* are to be found throughout the sub-continent. Their presence complicates further an already complex ethno-linguistic situation and often makes the task of modern nation-building in this part of Africa all the more difficult. There are several small Nguni (Ngoni) groups in modern Tanzania and Malawi. A large Nguni group occupies the area divided by the three boundary lines between the modern states of Malawi and Zambia, Malawi and Mozambique, and Mozambique and Zambia. In western Zambia the Kololo who befriended David Livingstone were an Nguni group whose influence is still evident, and Livingstone himself contributed to the Nguni diaspora by settling some Kololo in southern Malawi. At one time in the early 1970s two members of the Zambian cabinet had the surname Zulu. In Mozambique the Gaza people are part of the shattered remants of the defeated Ndwande.

Africa had a long history before European colonization. This is often forgotten because that colonization touched all parts of the continent with an often misleading impression of completeness. It is almost like a mantle of drift-deposits masking the underlying solid geology of the continent. The events of pre-colonial African history are most relevant to the modern political geography of the continent. Some modern states, such as Lesotho and Swaziland, represent nations forged in the pre-colonial *Mfecane*. Others such as Ethiopia and Morocco are of even longer lineage. Some pre-colonial African 'nations', such as the Ashanti, the Baganda and the Ndebele, play important roles as cultural and political minorities within modern African states and materially affect the well-being of the states of which they are now part. Above all it is necessary not to commit the error made by many of assuming that European colonialism wiped the slate clean in Africa and that nothing of what went before survived to have any significance to modern Africa. Great though its immediate impact has been, European colonialism, encapsulated within a single century, has to be placed properly in the context of a very long, complex and still influential pre-colonial African history.

# 3

# EUROPEANS AND AFRICA, 1415–1885

In August 1415 Ceuta, a fortified emporium on the north coast of Morocco, was taken by a Portuguese army which included the young Prince Henry the Navigator, who was to play a significant, though probably exaggerated, role in the Portuguese exploration of Africa. Ceuta was successfully held by the Portuguese until in 1593 it became Spanish when the crowns of Spain and Portugal were united. Ceuta remained Spanish after the separation of the two kingdoms in 1640, and is Spanish to this very day. Europe's first toe-hold is also Europe's last finger-hold on the African continent.

The taking of Ceuta is significant in the Christian fight-back against the spread of Islam under the Moors who had crossed into Iberia in the eighth century AD. Portugal, first recognized as a Christian kingdom in the twelfth century, enhanced its Christian status by taking the war to the Moors but also sought an economic future. The taking of Ceuta was not just a nationalistic and religious crusade but an attempt to secure the Straits of Gibraltar for Portuguese trade. It was also a bid for a stake in trans-Saharan trade because Ceuta was at the northern end of an important caravan route. The Portuguese would already have known of the fabled wealth of West Africa from reports of the pilgrimage of Mansa Musa of ancient Mali to Mecca via Cairo in 1347 when his camels were allegedly loaded with 15 tons of gold. West African gold encouraged the Portuguese voyages of exploration and permeated the European consciousness to the extent that Shakespeare could 'speak of Africa and golden joys' (King Henry IV, Part II, Act 5, Scene 3, Line 101)

Early European interest in Africa was founded on two interwoven motives, which were to prevail for almost five hundred years: Christian militancy and trade. In the late nineteenth century David Livingstone claimed that opportunities for Christian missions and 'legitimate' (non-slave) trade were the main attractions of the African interior.

The Portuguese drive along the west coast of Africa was aimed at the East and at securing the India trade then controlled by the unholy alliance of Saracens and Venetians in the eastern Mediterranean. The Portuguese were acutely aware of the richness of the prize but not of the great distance they would have to go to achieve it. Meanwhile the possibilities for trade along the African coast were themselves attractive and there were strong hopes of being able to sail up a river to the gold-rich Mali of Mansa Musa.

The Portuguese knew of the existence of the Christian kingdom of Ethiopia, ruled by the legendary Prester John. Ambassadors from Ethiopia reached Aragon in 1427 and Lisbon in 1452. The Portuguese conceived of a strategic Christian link-up with Prester John to outflank Islam. To these incentives must be added the scientific curiosity of the age which the Portuguese certainly possessed, the strong desire to find out what lay beyond. These several motivations led to one of the great episodes of world exploration.

Arab maps marked the north-west African coast as far as Cape Bojador, reached by the Portuguese in 1434. Between 1419 and the death of Prince Henry the Navigator in 1460 thirty-five expeditions left Portugal, eight initiated by Henry. Cape Blanco was reached in 1441 and Cape Verde in 1444, the year in which African slaves were first traded in Lisbon. By 1460 the three great European trades with West Africa had been established: gold, ivory and slaves.

Until the Portuguese pioneered the sea route from Europe to West Africa an important trans-Saharan trade flourished. Caravan routes crossed the desert from the Moroccan, Algerian and Libyan coasts to the Niger. Gold, ivory and slaves were traded for wheat, weapons and cloth. From the Sahara itself came salt from the mines of Taoudenni. Trading centres such as Timbuctoo and Gao flourished at ancient cross-roads where the east–west route along the river and the open yet watered savannah country met north–south Sahara caravan routes and routes from the tropical forest regions near the West African coast. Timbuctoo was a pre-industrial trading centre receiving goods from diverse regions, and a golden city, a centre of Islamic culture and learning and part of the Mali empire of Mansa Musa. Unbeknown to the Portuguese Timbuctoo was already in decline. In 1393 it had been plundered by a Moroccan army from across the Sahara and was now tributary to Morocco. The Portuguese trading posts on the West African coast contributed to its decline by attracting the trade of the West African forests away from the Niger. How far the Portuguese were responsible for the decline of trans-Saharan trade is a matter of debate, but gold, slaves and ivory were brought to their coastal stations in abundance. One fortified trading post on what became known as the Gold Coast, now Ghana, was named by the Portuguese *El Mina* (*São Jorge de Mina*, 1482). This reorientation of African trade routes, whatever its extent, was the beginning of European interference in African trade. The phenomenon greatly increased during the colonial period and in the post-independence era still has a profound and deleterious effect on the economies of modern Africa.

The Portuguese pressed along the coast of Africa seeking a sea route to India. Formal trade leases were made conditional upon further exploration of the coastline. Four great voyages of exploration secured the sea route to India. Each built on the experience and information carefully accumulated by its predecessor, even to the extent of changing the type of ship employed. They used as staging-posts the trading posts already established along the African coast. Diogo Cão left Portugal in 1482, found the mouth of the Congo (Zaire) river, and sailed on further south as far as *Cabo do Lobo* (Cape Seal) in August 1483, marking his progress by erecting stone *padroes* (crosses). In 1485 Cão returned to the Congo river, sailed up it to the head of sea navigation above present-day Matadi and then proceeded southwards to *Cabo do Padrão* (Cape Cross) on the Namib coast in January 1486. He probably died a short distance to the south having extended European knowledge of the African coastline by about 1,750 miles (2,800km) and explored the Congo estuary.

Bartolomeu Dias left Portugal in 1487 to follow Cão's route. He stood out to sea to avoid the worst of the Benguela current and rounded the Cape of Good Hope without sighting it. Dias followed the coastline of South Africa eastwards for about 450 miles (720km) before, in March 1488, erecting a stone *padrão* at Kwaaihoek. A little further east, at the Keiskamma river Dias turned for home (rightly) confident of having solved the problem of the route to India.

The actual opening of the route took another ten years. During that time the Portuguese, dismayed by the now obvious great length of the sea route, awaited news of their emissaries to India via the Red Sea. They also came to an accomodation with Spain over the New World at the Treaty of Toredsillas in 1494. An overland expedition, led by Paiva and Pero da Covilha, left Portugal in 1487 to try to reach India and Ethiopia via Cairo. Paiva died in Ethiopia but Pero reached India, explored the north-western littoral of the Indian Ocean and then made his way to Ethiopia. He established himself so well at the court of the Prester John that he was prevented from leaving and was still there when in 1520 a sea-borne Portuguese expedition arrived.

Some information from the da Covilha brothers got back to Portugal before Vasco da Gama sailed in 1497 to open the sea route to India with a fleet of four ships, two of which were the new square-rigged type recommended by Dias. Beyond Kwaaihoek da Gama had a long fetch of uncharted waters before reaching Arab trading posts on the East African coast where pilots could be obtained. They called at Quelimane, Mozambique, where they beat off an attack, Pemba, Mombasa, where they were again attacked because they bombarded Mozambique, and finally Malindi. Here *because* of their *contretemps* at Mombasa, news of which had preceded them, they were warmly welcomed. Having rested, revictualled and erected a stone *padrão* the Portuguese engaged a Gujerati pilot to take them across the Indian Ocean, arriving near Calicut on 20 May 1498.

The sea route to India was open. It was long and very dangerous. Thirty Portuguese ships were wrecked on the East India trade between 1498 and

1510 alone, most in the Indian Ocean. The route needed costly staging posts for revictualling and running repairs. These posts and the ships had to be protected, not only from Africans, but from the rival Arab traders and from the Dutch, French and British. As early as 1508 one Portuguese ship was captured by a French corsair in the Mozambique Channel. The Cape sea route became vital to the rival interests of the sea-faring nation-states of Western Europe, so much so that its protection became a major obsession.

The Portuguese tried to effect a strategic Christian link-up with Prester John in 1520 by an expedition via Massawa to the court of the Ethiopian Negus. Portuguese hopes of a strong, well-placed Christian ally to outflank Islam were dashed. Ethiopia was weaker than anticipated and sorely pressed by superior Muslim forces. It needed Portuguese military support to survive. Christoval da Gama, son of Vasco, led a small expeditionary force which from 1541 fought a long campaign against great odds. Da Gama was put to death after a pitched battle but Portuguese troops played a vital role in defeating the Muslims in 1543. From this time of early contact the Ethiopians, alone among Africans, were treated by the Europeans as equals. Helped by a resurgence of Ethiopian power in the nineteenth century, Ethiopia was again regarded by the European imperial powers as one of them, to the lasting discomfiture of her neighbours in the Horn of Africa.

West African slaves had been traded in Lisbon from 1444 but the trade burgeoned after Portugal and other European powers colonized the New World. Brazil had been 'discovered' in 1500 by a Portuguese East India fleet sailing westwards in order to exploit the currents and winds of the south Atlantic. As it was to the east of the line agreed with Spain in 1494 Portugal claimed Brazil as a colony. Plantations in Brazil became keen recipients of slaves from Angola and other parts of West Africa. The trade grew steadily as Luanda, São Tomé and Santiago became great slave entrepôts envied by rival European powers.

For Portugal was not alone in the slave trade. The Danes, Dutch, English, French, Germans and Spaniards also indulged. Brazil, the West Indies and North America were under-populated and unable to supply the labour required for the development of tropical and sub-tropical plantations of sugar, cotton and tobacco. The intensive labour inputs were provided by African slaves. A triangular trade developed between Europe, West Africa and the Americas. Cheap manufactured goods from Europe were traded in West Africa for slaves who became the human cargo for the notorious 'middle passage' to the Americas. The proceeds from their sale were invested in cargoes of plantation products to sell in Europe at great profit. A small proportion was then used to purchase cheap manufactures to start the cycle over again.

Another slave trade was conducted in East Africa by Arabs from the Swahili coast. Their slave entrepôts were Zanzibar and Socotra, and their markets the Persian Gulf, Arabia and India. Unlike their European counterparts in West Africa, Arab slave traders penetrated the African hinterland, carrying the Swahili language deep into the interior where it still survives in occasional place-names such as Bwana Mkubwa in Zambia. Their trading routes, surviving into the second half of the nineteenth century, were used by European explorers, for example, from Bagamoyo to Ujiji.

The explorers graphically described the Arab slave trade, its routes and fortified staging-posts. The horrors recounted confirmed that Britain had been right to ban the slave trade as early as 1807, and slavery itself in 1834. The anti-slavery movement had acted on humanitarian grounds based in late eighteenth-century enlightenment but in an economic context which favoured abolition. Wide areas of Africa were depopulated by slave traders, and whole communities were ruthlessly destroyed. Village was set against village, tribe against tribe and whole districts were laid waste. Once captured, slaves were forced to march, often over vast distances, to the coast, shackled together with chains or wooden halters. Many died on the march or on horrific sea voyages. Those who survived faced a lifetime of slavery in often appalling conditions before, all too often, an early death.

It is difficult to exaggerate the significance of the slave trades on Africa. They contributed to economic development in Europe, the Americas and the Middle East but systematically under-developed much of Africa. Angola, for example, still suffers from having been bled of its people and potential labour force for centuries. Slavery helped create racial antagonism and prejudice. Even where freed slaves were returned to West Africa from the Americas, problems were unwittingly created. Both Liberia and Sierra Leone have distinctive minority creole communities whose relative sophistication gave them political power only recently violently challenged by the indigenous peoples.

European rivalry, such a feature of the scramble for Africa, was manifest first over the East India trade. The sea route necessitated stations on the African coast and islands. These stations became the object of conflict between the European powers and a few also became bases for penetration of the African interior. At first the Portuguese were supreme. They captured Arab strongholds on the Swahili coast and built great stone island fortresses at Mozambique (Citadel of São Sebastian, 1558) and Mombasa (Fort Jesus, 1592), which still survive. For a time they held fief over all the other Arab cities of the coast: Sofala, Quelimane, Kilwa, Zanzibar, Pemba, Malindi, Lamu and Pate. On the west coast they built an stone island fortress at Luanda (Fort of São Miguel, 1576) and another on-shore at Benguela (São Felipe, 1587).

Portuguese supremacy on the route was seriously challenged by other west European nations from about 1600. East India companies were set up by the British (1600), Dutch (1602), French (1604) and Danes (1610).

The Dutch took Luanda from the Portuguese in 1641 and held it until 1648 when it was recaptured by forces from Brazil. The Dutch held St Helena in succession to the Portuguese until they founded Cape Town in 1652. The British occupied St Helena in 1659 and have held it ever since. Mauritius, discovered by the Portuguese, was settled by the Dutch from 1638 to 1710. In 1715 the French took over, changing the name to Ile de France. On 29 November 1810 the British, consolidating their grip on the sea route, took and held it until independence in 1968.

At the Cape the Dutch set up a victualling post and castle in 1652. Its strategic importance led the British to take it in 1795. In 1803 the Cape was returned to the Batavian Republic but when hostilities in Europe broke out again the British retook it in 1806. Confirmed as British in 1814 the Cape remained British until incorporated within the Union of South Africa in 1910.

Portuguese fortunes waned. They were thrown out of Mombasa by the Arabs in 1729 but held on to Mozambique. In the west they lost El Mina but regained Luanda and held the Cape Verde Islands. They largely lost out in the East and, partly as a result, their settlements on the African east coast became sleepy colonial backwaters.

In West Africa the slave trade was the most important, but gold, ivory, timber and beeswax, largely goods which were hunted or gathered, were also traded. European trading posts were strung along the West African coast like beads. Some were large and fortified, others were no more than the rude dwelling of a lone European. There was little penetration of the interior, and even the larger trading enterprises mostly depended on attracting trade to them. Traders sought quick profits but usually succumbed to diseases after a short tenure in their steamy, tropical outposts. Islands, which afforded greater security, were favoured locations, typically at river mouths, but often some way upstream on rivers navigable from the sea. For example, St Mary's Island at the mouth of the Gambia river, James Island about fifteen miles up the estuary and MacCarthy Island 160 miles from the sea all housed British trading posts, respectively Bathurst (Banjul), Fort James and MacCarthy (Georgetown). The trading posts represented little in the way of territorial claims and were to be found all along the West African coast from the French St Louis on an island at the mouth of the Senegal river to British Lagos, also on an island.

From the mid-nineteenth century the British and French governments attempted to rationalize their interspersed holdings on the West African coast but the proposals foundered in the face of opposition, in both countries, from commercial interests who wished to maintain the *status quo*. The deal fell through in 1876 before Germany claimed Togo (1884), so that had it been successful one large British colony might have developed where two British, one French and one German in fact evolved, later to become four independent states.

For the most part Africa was to Europeans an impediment on the sea route to India. Europeans rarely ventured into the interior. Portuguese penetration up the lower Zambesi followed the Arabs, taking over trading posts at Sena, Tete and Zumbo. They were far less successful than the Arabs in trading with the Monomatapa because their methods invited rebuff and gold supplies were running out. Little gold found its way to Lisbon compared with amounts earlier traded by the Arabs. Some Portuguese settlers took up land in Mozambique but numbers were small and most did not remain a distinctive group as they were far more relaxed about taking local African wives than the Dutch, French or British.

The Dutch at the Cape made the most significant pre-industrial penetration of the interior. In 1658 the Dutch East India Company allowed free burghers to acquire land. In time more settlers entered the colony, notably Huguenots after 1685. As land near Cape Town was taken up burghers began to drift eastwards. The dryness of the interior behind the Cape Fold mountains directed migration mainly along the coast. The pastoral economy practised was land-hungry and the sons of farmers (Boers) all expected farms of their own. The Khoi-khoi (Hottentots) and San (Bushmen) were pushed aside, sometimes by land purchase or conquest, often through decimation by European-introduced diseases such as measles and small-pox. The San were hunted down, for they had no respect for the white man's cattle and the Boer Commando system was born.

Before 1800 the Boers had reached the eastern Cape, over 400 miles east of Cape Town. Here they established the small town of Graaff Reinet (1786) but also encountered for the first time in numbers groups of Bantu-speaking peoples moving westwards along the same strip of land between the coast and the Great Escarpment. Their economy was similar to that of the Boers and equally land-hungry. The Bantu-speaking Africans were a much more formidable force than the Khoi-khoi, not least in numbers. The first major clash, known to the Boers as the 'First Kaffir War', to the Africans as the 'First War of Dispossession', and more neutrally as the 'First Frontier War', came in 1779.

The British came in 1795 to protect the strategic sea route. They captured Cape Town almost without realizing that the colony had acute frontier problems. They established Port Elizabeth in 1799 and during the short reign of the Batavian Republic Uitenhage, also near the frontier, was founded in 1804. Returning in 1806 the British faced a deteriorating frontier situation.

British rule, and that of the radical Batavian Republic, was not popular with the frontier Boers. Circuit courts, the abolition of the slave trade and interfering missionaries were resented. A Moravian mission had been established at Genadendal by Georg Schmidt in 1737 but he was refused re-entry to the Cape in 1743 because local Boers complained that 'to instruct the Hottentots would be injurious to the interests of the Colony' (Moffat 1842: 21). Johannes van der Kemp, a Dutchman who arrived in

South Africa in 1799 and under the auspices of the London Missionary Society established a mission at Bethelsdorp (1803), was the next missionary to trouble the Boers. Van der Kemp married a former slave girl and brought charges of murder and ill-treatment of slaves against several Boers. The capital charges failed but other convictions were obtained, to the lasting discontent of the Boers.

In 1815 another Boer myth of British injustice, that of Slaghter's Nek, was born when several Boers were hung as their 'rebellion' was harshly put down. The British were indecisive and inconsistent in their frontier policy. In 1820 5,000 British settlers, brought in to act as a buffer between Boer and Xhosa, failed to stabilize the frontier. Land taken from the Xhosa was returned by the Colonial Secretary Lord Glenelg who over-ruled the Governor of the Cape, an unwelcome early intrusion of imperial over settler interests.

The abolition of slavery in 1834 took Boers to the edge. The amount and method of payment of compensation pushed some over. The Cape was awarded in compensation less than one-half of the value placed on Cape slaves by the British themselves who then insisted that it was payable only in London. That the British taxpayer footed the bill for ending slavery did not count.

Some frontier Boers trekked away from British rule in the Great Trek which advanced the European frontier by over a thousand miles. The trek-Boers crossed out of the Cape colony onto the high veld. The African population, just recovering from the shock-waves of the *Mfecane*, tried to defend their land. Strongest resistance on the high veld was offered by the Ndebele under Mzilikase but they were finally defeated at Mosega in 1838. Retief led part of the trek into Natal where, with sixty-five men, he was treacherously murdered by the Zulu king Dingane in February 1838, but not before obtaining a treaty ceding Natal to the Boers. The Zulu made an indecisive attack on the Boer encampments but were defeated at Blood River on 16 December 1838. The Republic of Natalia was established with its capital Pietermaritzburg. On the high veld republics were set up at Wynburg and Potchefstroom, and later at Lydenburg and Ohrigstad. This early fragmentation was resolved into two republics centred on Bloemfontein and Pretoria, but the independently-minded Boers retained a tendency to spawn little republics from Stellaland and Goshen to Utrecht and Vryheid.

The British came by sea to annex Natalia in 1842. Many Boers trekked away from British rule for a second time. Although the Orange Free State and the Transvaal were at different times annexed by the British, the Boers were independent from 1881 to the end of the century. With British help they subjected Africans within the republics and pushed others to the margins to create future reservoirs of cheap labour, as in the mountains of present-day Lesotho. Slavery, by another name, was continued by the Boers who took African children as 'apprentices' in their frontier wars.

The Portuguese were attracted to Angola by slaves, souls and silver. Slaves became a vast trade; souls, from 1558, were the monopoly of the Jesuits; silver proved a false hope. Angola became dependent on Brazil as the main market for the slave trade which expanded greatly to the lasting detriment of the colony. Missionary work fell into decay and few Portuguese settled in the great, under-populated interior.

French contact with Senegal dated from 1483. During the seventeenth century the French government granted successive companies trading rights in ivory, gum, beeswax, hides and slaves. The first missionaries (Capuchins) entered Senegal in 1635, French settlement dated from 1638 and the town of St Louis from 1659. After 1815 attempts were made to introduce settlers but they collapsed after the withdrawal of the government subsidy in 1830. Climate, labour problems and African resistance were other reasons for failure. Thereafter the French confined themselves to trading for local commodities.

The French conquered coastal Algeria in 1830. *Colons* settled there, at first freely, then from 1840 under stricter control. Emigration from France was encouraged by free land, roads and planned settlements. *Colon* numbers increased steadily from 25,000 in 1839 to ten times that in 1871. Not all the settlers were French: many were Spanish, and others Italian, German, Maltese and Swiss. With the French these settlers demanded a say in the government of Algeria.

Missionaries often accompanied European thrusts into the interior. The trek-Boers, however, were not served by an ordained minister until the American Daniel Lindley answered their call in January 1840. Nor were there missionaries in significant numbers in Algeria until after 1869. Settlers and missionaries were not easy companions. As David Livingstone succinctly wrote in 1847: 'Boors [sic] hate missionaries' (Schapera 1961: 108). In southern Africa missionaries deliberately preceded the trek-Boers beyond the Cape colony boundary to reach Africans untainted by Europeans. Missions were established from Kuruman to Matabeleland, where a mission was finally achieved in December 1858 with the arrival of the Welsh missionary Morgan Thomas. The missionaries preached, learnt vernaculars and established orthographies for African languages for the first time. They could garden, farm and even print their vernacular testaments. In southern Africa they helped keep the road to the north open against Boer encroachment and provided the springboard and some of the personnel for the great explorations.

For centuries Europeans knew the coastline of Africa but not the interior. Climate, tropical diseases, Islam and resistant Africans deterred exploration. There was also a lack of interest, partly because trade at the coast was adequate and partly because there was little spirit of curiosity. In the late eighteenth century that changed. A spate of travel books excited interest. James Bruce journeyed to Ethiopia and the Blue Nile and belatedly told the story in five quarto volumes. Another Scot, Mungo Park, made two expeditions to the upper Niger but died at Bussa. He sailed past Timbuctoo which, along with the course of the Niger, became the object of European curiosity. By 1830, the mystery of the Niger, whether it was tributary to the Nile, was solved by Clapperton and the Landers but when Frenchman René Caille published his account of Timbuctoo in 1835 it was also not acceptable. Europeans did not want to read of the reality of mud houses and modest mosques but of a fabulous golden city. Heinrich Barth, a German commissioned by the Royal Geographical Society, finally laid the legend in 1855 in five stout volumes.

Big-game hunters and traders in ivory opened up the south but the outstanding contribution came from David Livingstone. In 1842–3 he made four separate journeys into the interior. By 1850, despite having his

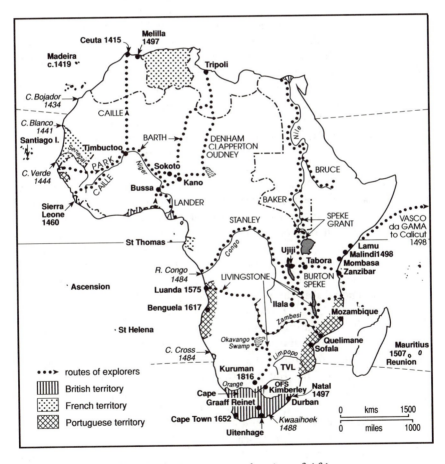

*Map 2* The European exploration of Africa

own mission station, he had twice crossed the Kalahari to Lake Ngami. In 1851 he first reached the Zambesi. Then, after despatching his family to England he left Kuruman in December 1852 to attempt to meet the need identified by his father-in-law, Robert Moffat, in 1842:

> It is now quite time to look to the eastern and western coasts of the continent, and form a chain of stations, from either, or both, towards the centre; and establish Missionary Colonies on lakes, or at the sources of those rivers which fall into the ocean. The want of navigable rivers, and the dry and often desert countries to be passed in Southern Africa in order to reach our isolated stations, present grievous barriers to the work of civilisation.
>
> (Moffat 1842: 191)

Livingstone returned to the upper Zambesi and struck west to reach the coast at Luanda (May 1854). He retraced his footsteps, and followed the Zambesi to its mouth, arriving at Quelimane in May 1856. In November 1857 his great book *Missionary Travels* caught the mood of the British public. In 1858 he was off again leading the far less successful Zambesi expedition. The river proved not to be navigable, and his wife Mary died (1862). Nevertheless he opened up the Shire River and Lake Nyasa areas to missionary influence. Livingstone's last journeys, begun in 1866, were aimed against the Arab slave trade and to discover the source of the Nile. He cut across the Arab slave routes but was largely powerless to interfere, and when 'found' by Stanley at Ujiji was dependent for his own survival on Arab slavers.

The source of the Nile had been one of the great geographical mysteries from the time of Ptolemy. Where did the great river that flowed out of the desert come from? Bruce traced the Blue Nile to its Ethiopian source. The most imaginative proposition as to the source of the White Nile was that of the *Jerusalemganger* Boers who in the 1860s thought that a north-flowing tributary of the Limpopo was the Nile; a small Transvaal *dorp* still bears the name of Nylstroom (Nile stream). More seriously, Burton argued for Lake Tanganyika as the source, Speke for Lake Victoria, each backing his own 'discovery'. Baker travelled up the Nile and met Speke coming down but their combined efforts left room for doubt and argument continued. Livingstone and Stanley together scotched the Lake Tanganyika theory by proving that no river flowed north out of the lake. Livingstone nurtured the theory that the Nile rose in Lake Bangweulu and was really the Lualaba. In 1873 he died at Ilala having failed to put his theory to the test. Stanley in his epic 999-day journey from Zanzibar to Matadi disproved the Lualaba theory but opened up the continent for the next stage of Europe's affair with Africa, the scramble.

Many other Europeans explored the interior of Africa. Their motivation was the prospect of Christian missions, trade, mineral wealth and scientific curiosity. Imperialistic rivalry between the European powers provided additional motivation as the exploration phase merged into the scramble. Exploration was a necessary prelude to partition and as time went on it lost its innocence and became caught up with the inexorable march of European imperialism.

# 4

# THE EUROPEAN
# PARTITION OF AFRICA

The term 'scramble for Africa' conveys the haste with which the African continent was partitioned between the European powers, mainly in the brief period 1884–1914. The scramble began in November 1884 with the Berlin Conference on the partition of Africa, called by the German Chancellor, Bismarck, and ended with the outbreak of the First World War. By then the initial partition was complete. The political map of Africa in 1914 closely resembles the modern political map.

At Berlin in 1884–5 the European powers agreed among themselves the ground rules for the partition, but in many ways the year 1869 was more significant in terms of the interest of the European powers in the continent of Africa. In that year there occurred momentous events at both ends of the continent to transform Europe's interest in Africa. In South Africa a diamond-rush started, with diggers attracted from all over the world to alluvial sites along the Orange and Vaal rivers. In the following year the fabulously rich 'dry-diggings' of Kimberley were discovered. The long-anticipated mineral wealth of Africa became a reality and a powerful incentive for the colonization of Africa was uncovered. In Egypt in November 1869, with great pomp and ceremony, in the presence of the Empress Eugénie of France (but alas without the première of Verdi's *Aida*), the Suez Canal was opened, affording a new direct sea route between Europe and the East. The strategic importance of Egypt, long appreciated from the ancient empires to Napoleon, was greatly enhanced. The diamond discoveries and the opening of the canal were important events in themselves but they also provided new starting points in Europe's interest in Africa. On the one hand diamonds represented Western capital's first major penetration of Africa and the start of industrialization, whilst the opening of the canal was soon to reawaken British imperial concerns over safeguarding the route to India against the rivalry of other European powers. The Franco-Prussian war of 1870 ended in defeat for France and was followed by the unification of Germany in 1872. These European events had far-reaching consequences for Africa: the humiliated French sought solace in creating a new overseas, mainly African, empire; the victorious

Germans sought colonies, also mainly in Africa, fitting to their new status of a unified empire.

The story of the discovery of diamonds in South Africa is well known. The salient facts are that the resource was extremely rich even on world-wide, all-time scales; it was highly localized – all four diamond pipes at Kimberley were within a 2-mile radius; they were discovered within an eighteen-month time-span; and Kimberley was at the heart of the southern African sub-continent, at least 500 miles (800km) from the ocean. The size of the resource and its localization made Kimberley a sharply focused, outstanding node of attraction. Its continental location ensured the opening up of an interior hitherto unrewarding and difficult for Europeans. Although the diamond diggings were located outside the British sphere, north of the Orange River, on farms recently allocated to its burghers by the Boer Republic of the Orange Free State, the diamond-fields were annexed by Britain ostensibly in support of the dubious territorial claims of a Griqua chief, Waterboer; in reality because: 'There was a notion also that the finest diamond mine in the world ought not to be lost to the British empire . . . [it was a] transaction, perhaps the most discreditable in the annals of English colonial history' (Froude 1886: 40). The boundary of the Orange Free State was realigned to pass less than a mile to the east of the diamond diggings. In 1876 the British colonial secretary, the Earl of Carnarvon, gave the Boer Republic an indemnity of £90,000 for relinquishing 'all claim to contraverted territory', which he subsequently asserted was: 'a sum which no one who considers that about £4,000,000 worth of diamonds are now annually extracted from the mines will say was an extravagant price' (Molyneaux 1903: 525). Diamonds transformed the South African scene. Railways, which reached Kimberley in November 1885, opened up the South African interior for the first time, provided a modern infrastructure and were the vehicle for political expansionism. The diamond diggings created large-scale local capital combined with important entrepreneurial and technical skills. Together they provided the springboard for further mineral exploitation which was to open up the sub-continent. Imperialism and capitalism were beginning to march hand in hand.

In 1886 gold was discovered on the Witwatersrand, which quickly became the greatest gold-field in the world. The nature of the resource, a few ounces of pure gold per ton of hard quartz, and its occurrence, in steeply pitched veins which quickly ran to considerable depth, meant that this was not a gold-field for the small prospector but rather for the heavily capitalized mining magnate. Local capital generated at Kimberley by Rhodes, Barnato and others quickly moved in to finance the development of the gold-fields. Whilst capitalism found political boundaries no impediment, this time imperialism did. The gold-fields, at the heart of the South African Republic (Transvaal), were too distant from British territory to be amenable to boundary 'adjustment'. Harsher methods were needed.

In December 1895, Dr Jameson, Rhodes' right-hand man, led a force into the Transvaal, hoping to be greeted by an armed uprising against the Pretoria government by the *Uitlanders* of the Witwatersrand. The raid was a total failure. There was no uprising and the Boers rounded up the raiders. Rhodes fell as Prime Minister of the Cape and Chamberlain narrowly retained his position as Colonial Secretary in London.

In October 1899 the British made a more determined attempt to take the gold-fields. In the conventional Anglo-Boer war, after enduring sieges at Ladysmith, Kimberley and Mafeking, the British eliminated the Boer army at Paardeburg to occupy Pretoria and force the Boer leader, Kruger, into exile. There followed a protracted Boer guerilla campaign and it was not until the 31 May 1902 that the two Boer republics and, more importantly, the gold-fields, came formally under British rule. The war, in which capitalism and imperialism were closely linked, gave rise through the works of Hobson and later Lenin to the theory that imperialism was the highest form of capitalism.

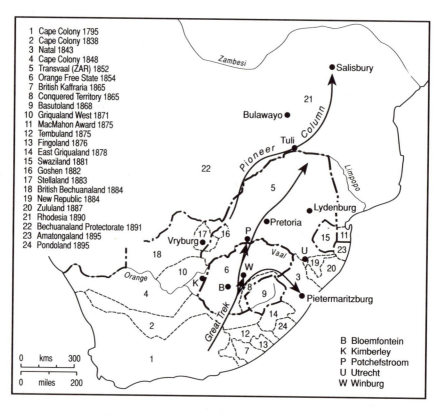

*Map 3* The expansion of European rule in southern Africa

Meanwhile the British, largely in the person of Rhodes, spurred on and financed by success in the diamond – and gold-fields, struck north. Having obtained a Royal Charter from Britain (1889) on the basis of doubtful mineral concessions from Lobengula, king of the Ndebele, Rhodes sent a pioneer column overland to Mashonaland, founding Salisbury (Harare) in September 1890. The British then took on and defeated Lobengula in 1893. A 'rebellion' in Matabeleland in 1896 was ruthlessly suppressed and the British South Africa Company ruled all Southern Rhodesia. It proved to be no Witwatersrand, as much gold had been extracted by Africans over centuries and traded via Sofala with the Arabs, and later the Portuguese. In the relative absence of minerals the Chartered Company turned to alienating land for white settlers as the basis of the new colony's prosperity.

Further mineral discoveries in southern Africa, whilst never again on the scale of Kimberley or the Witwatersrand, ensured that the whole of southern Africa was penetrated by Western capitalism and attendant imperialism. From Cape Town to Elisabethville the 2,300-mile (3,680km) spinal railway opened up the sub-continent, connecting a remarkable series of mining developments of diamonds, gold, coal, iron ore, zinc, lead and copper.

The modern Suez Canal had been built by the Frenchman Ferdinand de Lesseps. Britain had held aloof from the project despite overtures from de Lesseps until in 1875 Disraeli, with characteristic imperialist opportunism, purchased for £4 million the shares in the Suez Canal Company held by the impecunious Khedive of Egypt. The following year Britain and France assumed joint control of Egypt's perilous finances, sparking a nationalistic response from Egyptian army officers. The British government, now under Gladstone, had an investment to protect, and with some reluctance intervened to put down the Arabi revolt of 1882. That model of a modern major-general, Wolsely, further enhanced his reputation, earned mainly in little colonial wars, by winning the battle of Tel-el-Kebir. Britain took over the administration of Egypt, including the army, providing a nucleus of staff officers holding Egyptian titles. France, having declined to join the military intervention, faded from the Egyptian scene somewhat miffed.

The emergence of a unified Germany in 1872 brought another player onto the African stage. Individual traders such as Luderitz from Bremen appealed for German imperial protection. Bismarck was not immediately convinced of the arguments but was eventually converted. A less predictable entrant to the African scene was Leopold II, King of the Belgians, who aimed to carve a sovereign state out of the Congo basin under his International Association of the Congo. The position of Portugal was resented by the newer, more aggressive and more vigorous powers, though it suited Britain to support its 'oldest ally' because its admittedly tenuous but legal (in terms of the other European powers) hold on long stretches of the African coast kept out other more ambitious powers.

Bismarck, newly converted to colonialism, sought to bring order to the confusion of European interests developing in Africa and perhaps thereby to gain over his rivals. He called the European powers (plus the United States of America) to Berlin in November 1884 to draw up rules to regulate the partition of Africa. 'The General Act of the Conference of Berlin' was signed on 25 February 1885 by all the powers represented except the United States. Its expressed concerns were: 'the development of trade and civilization in Africa; the free navigation of the Rivers Congo, Niger, &c; the suppression of the slave trade by sea and land; the occupation of territory on the African coasts.' The last, relating to the partition, was elaborated as: 'being desirous . . . to obviate the misunderstanding and disputes which might in future arise from new acts of occupation (*prises de possession*) on the coast of Africa.' Any new claims to territory had to be notified: 'to the other signatory powers of the Act in order to enable them, if need be, to make good any claims of their own' (Hertslet 1909: 468–87).

The main purpose appears to have been to prevent conflict between the European powers themselves by establishing a set of accepted rules. African rights are not mentioned, except where they had been signed away to European powers. Otherwise they were simply ignored. The 'great game of scramble' was essentially a European game played to rules drawn up and agreed by the European powers.

Africa became Europe's geo-political chess-board. At the centre was the 'Independent State of the Congo' formed in August 1885 with Leopold as *Souverain*. The original concept of including the whole of the Congo basin was whittled down when the prior claims of the other powers were negotiated, but the Congo did retain a narrow corridor of access to the sea at the expense of Portugal, which retained Cabinda as a small exclave of Angola north of the Congo mouth.

Britons dreamt of the Cape-to-Cairo railway to be built with imperial, as well as commercial, considerations through territories coloured red on the map. It was the dream of a north-south axis over 6,000 miles (9,600km) long. Most of the railway built was financed commercially, mainly by mining, though some sections were constructed with strategic and military purposes as the main motives. In West Africa British colonial aspirations were not unified and eventually four separate colonial territories emerged, the largest being Nigeria in the east, the smallest the Gambia in the west.

The French and the Germans also thought of geo-political thrusts and axes and how best to counter British imperial ambitions. From Dar es Salaam the Germans successfully struck west to Lake Tanganyika (which was also the agreed boundary of the Congo) to prevent the British joining the two ends of the railway advancing from the Cape and Cairo. The Germans consolidated their position by building their own railway from the East African coast, eventually to reach Lake Tanganyika just before the outbreak of the First World War in 1914.

Rhodes, who thought geo-politically, aimed to keep the Boer republics from the sea. He also feared an east–west link between the Transvaal Boers and the Germans. In 1884 Germany established a protectorate over South West Africa whilst Kruger pushed the boundaries of the Transvaal westward and encouraged the small Boer republics of Goshen and Stellaland astride Rhodes' all-important 'road to the north'. Imperial force was needed to clear away the wayward Boers to keep the road to the north open and the Transvaal encircled. Containment of the Boers was not completed until 1897 when Britain annexed Tongaland. Further north on the east coast Rhodes himself was thwarted by the Portuguese from gaining access to the sea at Beira. British imperial interests backed the Portuguese even against the colonial Rhodes in order to keep the other European powers at bay by upholding Portugal's historic but otherwise shaky claims to long African

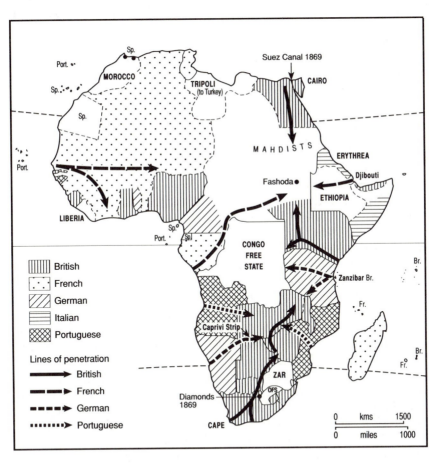

*Map 4* The geo-politics of the scramble for Africa

littorals. In 1890 Germany insisted on having access, by the Caprivi strip, to the Zambesi, which was patently not navigable. Of no immediate value, the strip might best be seen as a geo-political finger pointing from German South West Africa towards German East Africa.

In West Africa the French drove eastward from Senegal, to the Niger river, navigable for much of its length in the interior, and with a fertile inland delta. The early French colonial capital of Senegal was St Louis, the main port the present capital city Dakar. These two colonial towns were connected by the first railway built in tropical Africa, begun in 1882, over a distance of about 170 miles (270km). From St Louis the Senegal river is navigable as far Kayes in present-day Mali, a distance of about 465 miles (745km). In the 1890s the up-river voyage took eight days and twice-weekly steamers were scheduled. From Kayes the French built a metre-gauge railway over the 340 miles (550km) to Bamako on the inland navigable Niger. The French then advanced eastwards along the Niger, initially by gun-boat and army columns to take Timbuctoo in December 1894. Strange to relate: 'that Timbuctoo, a town nearly eight hundred miles from the sea – a town of the Sahara, moreover – was taken by sailors' (Du Bois 1897: 358). What is more, the immediate force of sailors numbered only nineteen, of whom seven were French. By advancing along the line of the Niger the French cut off any possible attempt by the British or the Germans to thrust too far inland from the Gulf of Guinea. The French were also acutely aware of British interest in reaching the Sahel from the north, from Morocco or Tripoli, though the reality of that threat is doubtful. The French effected a link-up with their advance from the Mediterranean coast of Algeria and so gained imperial control, albeit often tenuous, of much of the Sahara desert. Their main concern was to keep other European powers out of the area and, having accomplished that, they took their time to establish colonial rule over the vast and largely empty spaces of the Sahara where the local Tuareg offered dogged resistance to French military occupation.

Rivalry between the European powers led to several confrontations deep in the heart of Africa, for example, between the two well-known protagonists Sir Harry Johnston and the Portuguese Major Serpo Pinto on the banks of the Shire river in 1889. But the final act of the scramble for Africa in the nineteenth century was the near comic-opera incident at Fashoda when British hegemony over the Nile valley, having been challenged by the Sudanese, was tested by the French.

Egypt had taken possession of the Sudan in the early 1820s and their administration was taken over by the British who placed British and other European officers in charge of the regions. In 1881 the revolt of the Mahdi began against Egyptian rule and quickly spread throughout the Sudan. The Egyptians and their British officers were driven out in a remarkable wave of fervoured nationalism which culminated in the defeat and death of General Gordon at Khartoum in 1885. In Equatoria Province the

Governor, Emin Pasha, was forced to flee the Mahdi's forces and was reluctantly rescued by the intrepid Stanley in 1889 on his last epic trans-Africa expedition which bridged the exploration and scramble phases of Europe's torrid affair with Africa. Only in 1898 did an Anglo-Egyptian army under the *Sirdar* (Commander-in-Chief) Kitchener advance up the Nile to avenge Gordon and finally to secure the valley of the Nile for Britain. The large Anglo-Egyptian army advanced slowly up the Nile, methodically building railways and launching river-boats to secure the long supply lines. In September 1898 Kitchener met and defeated the forces of the Mahdi at the set-piece battle of Omdurman, and continued on up the Nile with his vast army.

Meanwhile a small French force of eight officers, three NCOs and 130 Senegalese troops had begun a secret journey from the west coast to the upper Nile. There they planned to meet up with another French force advancing from the east coast at Djibouti with the help of the Ethiopians who had recently established their independence by defeating the Italians at Adowa in 1896. Under the command of Major Marchand the first force crossed from Libreville on the coast to Brazzaville on the Congo. Thence they sailed up the Congo and Oubangui rivers to the limit of navigation. Dismantling and dragging their boats through the watershed swamps they completed a difficult portage to the western headwaters of the Nile. They then sailed downstream to join the White Nile and camped at Fashoda on the left bank, over 400 miles (640km) above Khartoum/Omdurman, to await the arrival of the second French force from Djibouti.

Unbeknown to them that force met disaster in the shape of disease as it descended from the Ethiopian highlands to the hot Sudanic plains. Upstream instead came the great flotilla of the Anglo-Egyptian army under Kitchener. With pomp and courtesy on both sides, the niceties of international diplomacy were enacted under the scorching sun of the Sudan. All the cards were in Kitchener's hand; an army of thousands well equipped with field-guns fresh from a major victory facing a tiny French party. Moreover, Kitchener had complete control over the lines of communication, the telegraph lines carefully laid all the way from Cairo and from there to Europe. Messages buzzed along the wires but Major Marchand got little useful support from Paris where the government was in the turmoil of the Dreyfus affair. The charade of diplomacy was acted through to its inevitable end. Marchand and his little army abandoned their position, but with swords drawn and flags flying, even spurning the easy but potentially humiliating British-controlled route down the Nile, to march to Djibouti and return as heroes to France.

The scramble for this part of Africa was consequent upon ensuring the security of the British imperial route to India. Security of the route to India (until 1869 the Cape Sea Route) had been a main plank of British foreign policy for many years. When a new, shorter route became available through

Suez it was inevitable that the British would attempt to control it. But to control the Suez Canal it was necessary in crisis to occupy Egypt, which in turn, because the Nile was its life-blood, was secure only when the Sudan was safe. In East Africa the British and Germans had sorted out their differences and the British sphere included Uganda, the effective source of the Nile, which became a British protectorate in 1894.

Italy was a late-comer to the scramble for Africa, having been preoccupied with its own unification. Although anxious to get involved, partly, like Germany, to enhance its new-found status, not much of Africa was left unclaimed by the other European powers. In the Horn of Africa the part of Somalia not taken by the British came under Italian protection, and in 1890 Eritrea, on the Red Sea, was made an Italian colony. From Eritrea Italy resolved to take Ethiopia, to link Eritrea with Somaliland and so form a large and coherent Italian East African Empire.

The Italian army advanced into northern Ethiopia from Eritrea, was met at Adowa by the Ethiopian army on 1 March 1896 and was routed. Victory secured independence for Ethiopia which was thereafter treated by the European powers as an equal in the partition process. Although Italy eventually avenged Adowa by taking Ethiopia in 1935–6 to form the long-desired Italian East African Empire, that success lasted a mere five years. The short period of Italian rule did not prevent Ethiopia emerging from the Second World War as an imperial force itself to dominate the international politics of the region.

In October 1911 Italy invaded Tripoli, Turkey's one remaining *Vilayet* in Africa. In those pre-petroleum days Italian motives seemed to be that Tripoli was near, had been Roman and the nationalists thought it necessary 'to wipe out the shame of Adowa' (McCullagh 1912: 9). The adventure almost proved calamitous at the battle of Sharashett, but the Italians scraped victory in a controversial war amid allegations of massacres and atrocities.

Towards the end of the scramble for Africa, rivalry between the European powers increased. A major issue was that of free trade in all African colonies, an issue featured in the Final Act of the Berlin Conference in 1885. Morocco, the only African political entity other than Ethiopia to be treated as an equal by the European powers, being independent, was open for trade to all European powers. Germany built up a considerable commercial interest in the country with large exports, many merchants, nine Consulates, fourteen Post Offices and three shipping lines with over three hundred ships calling annually at Moroccan ports. France wished to make Morocco a French protectorate, and by separate treaties with Britain and Spain . . . both with secret clauses and respectively signed in April and October 1904, was given a free hand in Morocco. Getting wind of this development, in March 1905 the German Emperor visited Tangier to: 'make it known that I am determined to do all that is in my power to safeguard efficaciously the interests of Germany in Morocco. I look upon the Sultan as an absolutely independent sovereign' (Morel 1915: 75). A conference of European powers and Morocco was held at Algeciras in 1906 and confirmed: 'the independence of the Sultan, the integrity of his dominions and economic liberty without any inequality' (Morel 1915: 27–8). Hollow words. Although a signatory at Algeciras France continued to undermine Morocco's independence. Military intervention was made on a number of pretexts and Morocco slipped towards French 'protection'. In July 1911 Germany intervened by sending the small gunboat *Panther* to Agadir. The Agadir crisis, which was a European crisis anticipating the First World War, developed as it became known that Britain would support France in any potential conflict.

The situation was defused when France and Germany came to an agreement on 4 November 1911. Under that convention Morocco was partitioned between France and Spain but with freedom of trade for all European powers. In compensation France ceded to Germany 107,270 square miles (277,829 square km) of the French Congo adjacent to Kamerun with access to the Congo river by two corridors (shades of the Caprivi strip). By way of *quid pro quo* Germany ceded to France 6,450 square miles (16,705 square km) of upper Kamerun which became part of Chad. This curious territorial settlement did not long grace the political map of Africa because the Versailles Treaty at the end of the First World War restored the *status quo ante*, apart from the gain to Chad.

To the European powers the partition of Africa became a gigantic game, some super 'Monopoly', played with real land and real people. Thus Britain traded the North Sea island of Heligoland with the Germans for Zanzibar and parts of northern Nigeria with the French for fishing rights off Newfoundland. In 1924 the British moved the boundaries of Kenya westward, to give Jubaland to Italy as a reward for its support in the First World War. France ceded part of Chad to Italy for the same reason in 1935 but the Rome or Laval/Mussolini Treaty was not fully ratified, thus becoming a cause of friction between Chad and Libya, resolved by the International Court of Justice only in 1994. French face at Fashoda was in part saved by Britain agreeing to give the French 'a free hand in Morocco', as confirmed in the 1904 *Entente Cordiale*. The Franco–German Convention of 4 November 1911 exchanged land in equatorial Africa for the same 'free hand in Morocco'.

Africans were not disinterested bystanders, nor reluctant but inactive victims of European political aggression. There was resistance, which in the first centuries of the European experience of Africa played a large part in deterring European penetration of the continental interior, and in the nineteenth century, when stronger motives moved Europeans to venture inland in force to alienate land, to exploit resources and to threaten local polities, resistance became even stronger. The ten frontier wars in the Cape lasted one hundred years (1779–1879). The trek-Boers faced stern and protracted resistance to their invasion of the South African high veld from the Ndebele led by Mzilikase, the Sotho led by Moshoeshoe and the Bapedi led by Sekokuni. Zulu power in Natal was overcome by the Boers at Blood River in 1838 and by the British victory at Ulundi in 1879, but there was still the Bambata 'rebellion' of 1906. It was not until the bloody wars of 1893 and 1896 that the Ndebele were finally crushed by the British in Matabeleland. The period of scramble is littered with the carnage of 'little wars' of colonial subjugation all over Africa, from the Ashanti Koffi in 1874 and Prempeh in 1895, Egypt led by Arabi in 1882, Sudan and the Mahdi in 1885, Ethiopia in 1896 to Libya in 1911. It was not for want of resolve to resist that almost all were won by Europe. It was the superior technology of rifle, field-gun, Gatling and Maxim machine-guns against assegai and obsolete muzzle-loaders; greater resources and numbers. Nevertheless there were occasional devastating European defeats, for example, for the English at Isandlhwana (1879) and the Italians at Adowa (1896). As the scramble developed, not only did the fire-power available to the Europeans increase, but so did their ruthlessness. The British in Matabeleland and Germans in South West Africa spoke of 'extermination' of groups of Africans less than a century ago. All over Africa there was resistance, often on a lower key from the set-piece battles above, and that resistance smouldered on well into this century.

Africa became a convenient venue for European powers playing out their rivalries, each asserting its perceived status, satisfying its national ego. In large part the fact that these rivalries were exercised in Africa was incidental. It might have been anywhere away from Europe and Africa happened to be the place chosen. When the European rivalries were focused in Europe itself, as they were in 1914, perhaps in part because there was nothing left of Africa to bicker over, the result was a devastating world war. Of course there was an element of economic exploitation among the motives for the scramble; mineral resources were plundered, cash-crop economies were implanted for the benefit of Europe and cheap manufactures were marketed to expand European industry. But although these factors were locally dominant with far-reaching effects for parts of Africa, taken as a whole they were secondary considerations in the minds of the European politicians responsible for the partition of Africa. Compared with Roman knowledge of British resources before 55 BC, European knowledge of African resources in 1885 was woefully small. Hope, not expectation, of mineral riches provided some motivation but the hope was not always fulfilled.

The European partition created fundamental problems in Africa that are still present and are likely to be there for many years to come. The way the essentially European rivalries were worked out affects the basic political structure of modern Africa. The division into over fifty colonial territories, with typically almost total disregard for African interests and a total indifference to the rights and wrongs of trading so casually in African territories and people, lives on even though those colonies are now independent sovereign states. The partition of Africa was by Europeans for Europeans. Its relevance is that the divisions imposed upon the continent became, with very little change, the geographic framework for African independence.

# 5

# COLONIAL AFRICA

The political map of colonial Africa was more or less complete by 1914. The ensuing First World War brought territorial changes but more importantly altered the status of some colonies in a way which was to be significant to the political development of the whole continent.

Because Germany had four colonies in Africa – East Africa, Kamerun, South West Africa and Togo – four colonial wars were fought in Africa and, in addition, the Turks threatened the Suez Canal from the east. Except in East Africa the wars were short and of little consequence. In Kamerun and Togo the Germans were defeated by the British and French from their neighbouring colonies of Nigeria and French Congo on the one hand, and Gold Coast and Dahomey on the other. In South West Africa the Germans were overcome by South African forces who, led by Generals Botha and Smuts, made a three-pronged attack from the ports of Walvis Bay and Luderitz Bay and overland from South Africa itself. In the overland advance the Cape rail network was extended to link with the German rail system which was upgraded to Cape gauge. In East Africa the German General Von Lettow Vorbeck fought a brilliant guerilla campaign against British and South African forces. As the only undefeated German General of the whole war he was fêted on his return to Berlin in 1919.

Between 1914 and 1919 Africa was effectively repartitioned. At Versailles Germany renounced all overseas possessions in favour of the League of Nations whose Covenant stated that the well-being and development of the peoples of the former German colonies: 'not yet able to stand by themselves ... formed a sacred trust of civilisation ... [and] securities for the performance of this trust should be embodied in this Covenant' (Wellington 1967: 262–3). Mandates to administer the former German colonies were issued. Kamerun and Togo were each divided between the British and the French. The British administered their portions with Nigeria and the Gold Coast respectively, whilst the French ruled theirs as separate units. German East Africa was divided; by far the smaller part, Ruanda-Urundi, contiguous with the Congo, went to Belgium, the other, Tanganyika, to Britain. South West Africa, because of its lack of development and contiguity, was to be

46

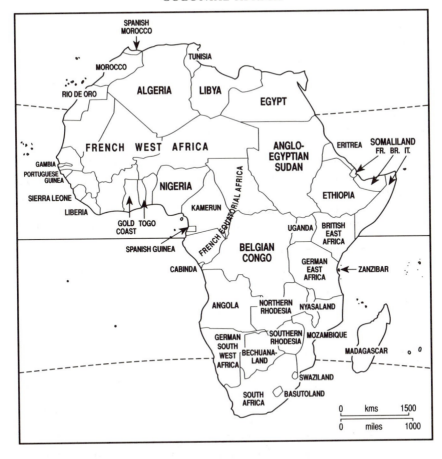

*Map 5* Africa in 1914

administered as an integral part of South Africa 'subject to safeguards in the interests of the indigenous population' (Wellington 1967: 263).

The League of Nations set basic standards for the mandated territories, for example, banning slavery, arms and alcoholic drink. An annual report from the mandatory powers on each territory was required by the Permanent Mandates Commission at Geneva. The Commission also welcomed petitions concerning individual territories. The mandate system caused light to be shed on African colonies and, whilst there was no coercion of the mandatory powers the publicity attending their administration was conducive to good practice. Furthermore the phrase 'not yet able to stand on their own feet' was widely interpreted as implying that the ultimate goal for the territories, though many years away, was self-rule and independence. In 1919 that was indeed a radical concept.

47

After the Second World War the League of Nations mandates were transferred to the United Nations (UN) and the system was extended to Somalia, the former colony of Italy. At this time the concept of colonial independence was not only made explicit but the UN actually laid down a timetable for independence in each territory. These were adhered to and the mandated territories generally set the pace for independence in Africa. Thus the territorial booty of the First World War became the Achilles heel of European colonialism in Africa. Only the mandate of South West Africa (Namibia) was troublesome because apartheid South Africa refused voluntarily to transfer its mandate to the UN and because the growth of apartheid was clearly not in the interests of the indigenous population. Far from setting the pace in its part of Africa Namibia achieved its independence only in 1990 after every other African territory except Eritrea (1993) and Western Sahara, and the advent of majority rule in South Africa (1994).

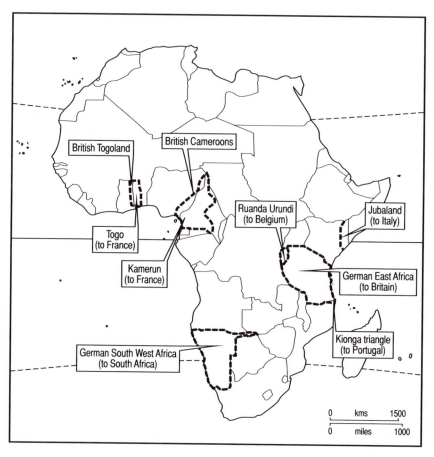

*Map 6* Changes in colonial Africa consequent upon the First World War

Some territorial changes resulted from the First World War. Portugal quietly took back the Kionga triangle seized by Germany in 1909. Britain honoured the secret Treaty of London (1915) and gave Italy Jubaland. The area ceded to Kamerun in 1911 returned to French Congo, but the part of Kamerun ceded to Chad under the same Convention was retained. The Caprivi strip disappeared until 1929, administered as part of Bechuanaland, but was then returned to South West Africa.

A major consideration during the scramble for Africa was that colonies represented markets for the manufactured goods of the metropolitan countries. Their mineral or cash crop raw material resources were also exploited to feed the same home manufacturing industries. The colonies were therefore, at least in theory, valuable assets. This was not always the case because the cost of colonial administration could be very high. For example, a colonial civil service, a judiciary and an enforcing military presence had to be paid for. The imperial powers saw that colonial administrations balanced their books to prevent colonies becoming a liability.

Out of this financial constraint in the British case came the concept of indirect rule, or in Lord Lugard's words 'the dual mandate' (Lugard 1929: title). The basic idea was to leave in place as far as possible traditional admin- istration and justice under an umbrella of low-cost British supervision. In this system British ex-patriot District Officers were spread thinly, covering vast areas and large populations, essentially by merely overseeing local traditional administration and dealing with problems that could not be readily solved at a local level. An administrative hierarchy was created for each colony from District Officer up to Governor.

There was also, in the British case, a hierarchy of colony and protectorate. The colony was usually at the coast, small in extent and more intensively ruled. Colony status also applied where there were white settlers. The protec- torate was often inland and more remote, thinly or indirectly ruled with few if any settlers. British rule in the protectorates was almost 'semi-detached'. Thus Lagos, Kenya and Southern Rhodesia were colonies, Nigeria, Uganda, Northern Rhodesia, Nyasaland, Bechuanaland, Basutoland and Swaziland were protectorates. The key to the whole system was making the colonial books balance.

The French operated a much more intensive colonial rule, perhaps more consistent with their concept of the superiority of French culture, which would not as easily allow for indirect rule as in the British case. The French equivalent of indirect rule was military administration which was practised for many years in remoter parts of their empire in Africa, especially in the Saharan territories. This too was thinly spread, relatively cheap and gave the French army the opportunity to be in a constant state of active duty. The French also kept ringing the administrative changes, the most notable of which was Upper Volta, a separate colony in 1919, which was divided between the Ivory Coast, Niger and French Sudan in 1932, to re-emerge as a separate colony in 1947.

Under both British and French different intensities of colonial administration applied to different parts of the colonial empires. This was matched by different levels of economic development, in terms of private and public investment and in infrastructural development. In part these differences were related to the pattern of resource distribution but there was a large distance decay factor from the main colonial administrative centre. So the more remote areas, unless they were heavily endowed with resources, were lightly administered, whether by indirect or military rule, and were less developed. But such areas were defined, if only for the less intensive form of colonial rule, and in due course they achieved full independence as separate colonial units, often as very poor, less developed, land-locked states.

Another form of indirect rule, employed by all European powers, was to lease out tracts of their African colonies or protectorates to private commercial companies. These companies were given monopolies, sometimes by royal charter, in resource production and trading and wide powers to administer all aspects of a territory, including, for example, the right to raise an armed police force and to tax the African population. The degree of supervision from the metropolitan country or an official colonial administration was usually minimal. There were also concessionary companies with more limited powers, for example, for construction and running of a railway line. The scale of operations varied from virtual control of an area the equivalent in size of several modern states to control of a very small territory or a single utility. The companies represented privatized colonialism and ensured the almost unfettered penetration of Africa by Western capital, with sometimes horrific effects on Africans subjected to the system.

Britain granted royal charters to companies engaged in many parts of Africa. Among the best known were the Royal Niger Company (1886) and the British South Africa Company (BSAC) chartered in 1889. The BSAC was given rights over 711,000 square miles (1,137,500 square km) of northern Bechuanaland and Southern and Northern Rhodesia, an area six times that of Great Britain. The royal charter was gained because the BSAC held a concession from Lobengula, king of the Matabele for:

> the complete and exclusive charge over all metals and minerals situated and contained in my kingdoms . . . together with full power to do all things . . . necessary to win and procure the same, and to hold, collect, and enjoy the profits and revenues, if any, derivable from the said metals and minerals.
>
> (Hiller 1949: 219)

This was the infamous Rudd Concession, signed on 30 October 1888. Lobengula almost immediately regretted signing it and tried in vain to repudiate the agreement, even sending two *indunas* as emissaries all the

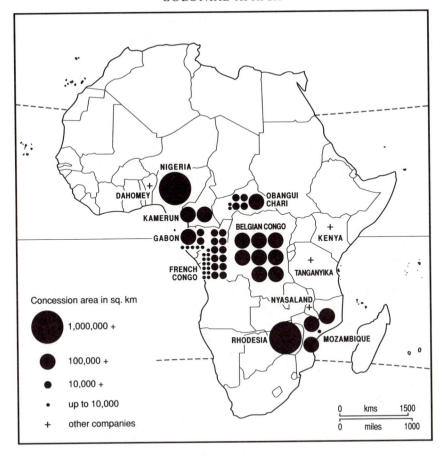

*Map 7* Chartered companies

way from Bulawayo to London where they had an audience with Queen
Victoria. They failed in their purpose against concerted opposition from
Cecil Rhodes supported by the British High Commissioner at the Cape,
Sir Henry Loch. The royal charter was granted and Rhodes sent his right-
hand man Dr Jameson to Matabeleland to gain Lobengula's agreement for
a column of white settlers, under the auspices of the BSAC, to enter
Mashonaland. In September 1890 the pioneer column founded Fort
Salisbury which was to become the local headquarters of the BSAC and
capital of Southern Rhodesia. There followed two wars (1893 and 1896)
in which imperial forces assisted the establishment of company control over
the Matabele by force of arms.

Long before the settlers entered Mashonaland, let alone before *de facto*
control was established the royal charter gave the BSAC wide-reaching

powers in return for: requiring the BSAC always to be British and under the final control of the Secretary of State for the Colonies; abolishing slavery and 'regulat[ing] the traffic in spirits and other intoxicating liquors'; and maintaining freedom of religion and the administration of justice with 'careful regard' to traditional customs and laws. The BSAC was empowered to establish or authorize banks and other companies 'of every description' to make and maintain roads, railways and telegraphs, to carry on mining and other industries, to make concessions of mineral and forestal rights (for a 50 per cent share of profits), to 'irrigate and cultivate any lands', to settle any such lands and 'to aid and promote immigration' (Eybers 1918: 559–66). The British government, through the crown, had the arrogance to grant a British company all these rights in someone else's country on the basis of the company owning a repudiated mining concession given in return for one hundred pounds sterling per lunar month, one thousand rifles with ammunition and a small gun-boat on the Zambesi to defend Matabeleland. The Rudd Concession, limited to mining rights, was bought by Rhodes for shares in the BSAC to the value of £1 million.

There was opposition to this hijacking of a country from British liberals, but it was not effective, any more than was the outcry in response to the harsh repression of the 'Matabele Campaign' of 1896 which prompted Olive Schreiner's powerful polemic *Trooper Peter Halket of Mashonaland*. The Rudd Concession was actually witnessed by English missionaries who, as advisers and interpreters to Lobengula, clearly favoured the infiltration of British capital.

*L'Etat Indépendant du Congo* in 1885 became the personal colony of King Leopold of the Belgians through the International Association which he created. Specifically mentioned in the Final Act of the Berlin Conference was that free trade should prevail throughout the Congo Basin. Trading companies of several nationalities established agencies in the Congo, trading for ivory and other products with the local people. Leopold meanwhile decreed that all land in the Congo apart from actual villages and the land cultivated by them belonged to the state, that is, to him. In 1891–2 three more decrees made it illegal for Africans to sell products of the forest to European traders, or for the traders to buy such products, on the grounds that the products themselves belonged to the state. Most in demand were rubber and ivory, the latter of course implying that the animals were also the property of the state. Leopold set up several concessionary companies with exclusive trading rights over defined areas and took a 50 per cent stake in each. Africans were taxed so that they had to gather produce from the forest for the concessionary companies almost as a form of tribute, and this was brutally enforced by the *Force Publique* and, in some cases, native levies which the companies themselves were authorized to raise and equip. In essence this was the so-called 'Congo system', a means of ruthlessly exploiting the resources and people of a colony.

In 1899 the same system of exploitation was introduced into the contiguous French Congo. Most of the colony was divided among concessionary companies, largely controlled by the same groups as were behind the companies operating in the Congo Free State. The products most traded were also rubber and ivory. Because in the French Congo the system was not enforced so ruthlessly it was less of a 'success'.

The Congo system attracted much contemporary criticism and opprobrium largely because of the way in which the private companies, backed by the official 'army' of the state, brutally abused and exploited Africans. Much of the criticism was aimed specifically at Leopold and this, in part, led to the Congo, which shared the same monarch as Belgium, becoming the Belgian Congo in 1908, that is, a colony of Belgium. The colony was almost one hundred times greater in extent than the metropolitan country. The abuses did not stop with the change in administrative status and there was evidence to show that Belgian ministers spoke in defence of the system, although the Belgian government was fully aware of the situation in the Congo. Contemporary critics alleged that, at a conservative estimate, ten million Africans died in the Congo as a result of the system. Some were brutalized by the armed levies but many more simply starved to death as they were coerced into spending all their time collecting rubber, for which they were paid a pittance, rather than producing their own food. By the First World War the Belgian and French Congos were exhausted colonies, stripped of their resources and pillaged by the system. Rubber from Asian plantations supplanted the wild Congo product and the system died. At the Treaty of Versailles, despite its shocking colonial record, Belgium was awarded the trust territory of Ruanda–Urundi under League of Nations mandate.

The Congo System was heavily criticized by many deeply concerned about the humanitarian aspects of the scandal. The British secretary of the international 'Congo Reform Association', E. D. Morel, was a prolific propagandist for the cause. The system, by ceding territory to monopolistic companies closed vast areas to free trade and so violated the Final Act of the Berlin Conference which had declared the Congo basin a free trade zone. More critically the System made a mockery of the Act's aim 'to instruct the natives and bring home to them the blessings of civilisation' (Hertslet 1909: 473).

Not all colonies were run in such an exploitative manner and not all concessionary companies were run on the same basis as those in the two Congos. Germany and Portugal also gave concessionary companies monopolistic rights over large areas. In Kamerun two such areas were alienated, whilst in Mozambique a very high proportion of the colony, was given over to three companies: Mozambique, Zambesia and Nyassa.

In most colonies administrations were keen on attracting private capital from the metropolitan country by giving concessions for railways, mining and other enterprises. For example, in the French colony of Dahomey (Benin) a railway was constructed from Cotonou, the colonial port-capital, inland for 377 miles (605km). Construction, begun on 1 May 1900, was shared between the colonial administration and a concessionary company. The colony undertook the construction of the permanent way, earthworks, embankments and cuttings, in fact everything but the laying of the track, sleepers, rails and rolling-stock which were the responsibility of the concessionaire. The colonial administration agreed to advance an annual sum of 1 million francs (£40,000) for five years towards the total cost of 11.7 million francs (£469,000) for the first section of 115 miles (186km). The labour needed for the construction work was considerable, up to 5,000 men, which local chiefs had to provide. The colony benefited from the increased trade tapped by the railway, the concessionary company from the railway revenues and profits. The metropolitan country made no formal contributions because government investment in the project was totally financed from within the colony.

The French were the first to organize their African colonies into federations. Such arrangements led to centralized hierarchies of administration in Africa which simplified working relationships between the metropolitan government and the colonies. Federation meant that an overview could be taken of the problems of highly disparate territories, some of which, although vast in area, were very sparsely populated and desperately poor in contrast to others which were quite densely populated and reasonably rich. Administration of the great open spaces of the Sahara and its fringe were in part underwritten by the more highly developed and richer coastal colonies.

The *Afrique Occidentale Française* (AOF) was formed in 1902 and the *Afrique Equatoriale Française* (AEF) in 1908. The former comprised eight colonial territories covering an area of 1.8 million square miles (4.7 million square km), the latter four territories of about 1 million square miles (2.5 million square km). The government of each was headed by a Governor-General, based in Dakar and Brazzaville respectively. Each individual territory had a colonial governor or military commander, and each colony was divided into several administrative districts. Federal control was taken for defence, both internal and external, and for trunk transport which included three railways which crossed the boundaries of individual colonies.

The British also created a federation of colonies in Africa, but very much later than the French and for different reasons. The Federation of Rhodesia and Nyasaland, which united the two Rhodesias and Nyasaland, was formed in 1953 by the Conservative government in London with the interests of the white settlers of Southern Rhodesia uppermost. There was opposition to the idea from African nationalists, particularly in the two northern territories where there were very few white settlers, and from a strong body of opinion in Britain itself. Extreme right-wing opinion in Southern

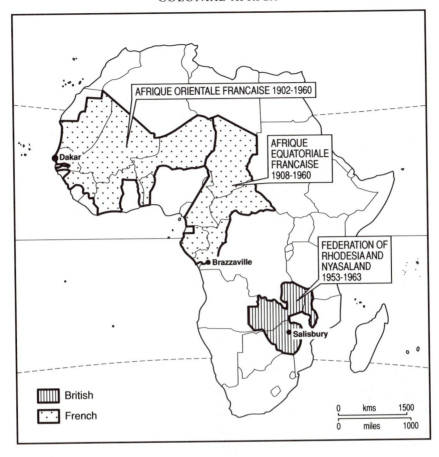

AFRIQUE ORIENTALE FRANCAISE 1902-1960

AFRIQUE
EQUATORIALE
FRANCAISE
1908-1960

FEDERATION OF
RHODESIA AND
NYASALAND
1953-1963

Dakar

Brazzaville

Salisbury

British

French

*Map 8* The colonial federations: AEF, AOF and Rhodesia and Nyasaland

Rhodesia was also against federation, not wanting to be burdened or threatened by the large black majorities of the northern territories.

Protagonists of federation hailed the whole enterprise as a bold new Elizabethan venture in positive racial partnership, and contrasted their vision with the strongly negative apartheid state which was being grimly built in neighbouring South Africa. They gave federation an economic rationale. A single economic unit of the three territories gave better prospects for industrial development through the creation of a common market whose size encouraged manufacturing industries needing a higher local market threshold. There were perceived complimentarities between the territories: the mines of the Northern Rhodesia Copperbelt were backed by the coal resources, the relatively highly developed industry and commerce of Southern Rhodesia, and the labour exports of rural Nyasaland. A federal government could take decisions, particularly over provision of modern infrastructure, to benefit all three components in this economic structure. Thus the Kariba hydro-electric dam built (from 1960) as a large federal project on the Zambesi river symbolically linked Northern and Southern Rhodesia and supplied electricity to the copper mines of the north and also to the towns and manufacturing industries of the south. A new railway line (1955), direct from Southern Rhodesia to Lourenço Marques (Maputo), pointedly avoiding South Africa, symbolized economic and political independence.

The Federation lasted for little over ten years or, in the title of Prime Minister Welensky's subsequent book, *4000 Days*. Its duration was entirely encapsulated within the 1951–64 period of Conservative rule in Britain. The immediate causes of break-up of the Federation were political. African political aspirations were not met and the reality of racial discrimination within the Federation mocked the flaunted ideal of 'racial partnership'. Nyasaland and Northern Rhodesian nationalists wanted to catch the tide of independence that flowed so strongly elsewhere in Africa. There was escalating political unrest in the northern territories and the British government bowed to the pressure. Within a year of the end of federation Nyasaland was independent as Malawi and Northern Rhodesia as Zambia. Among the underlying causes of break-up, economic factors were also important. The perceived complimentarities of federation did not translate into benefits for the northern territories, but did help the south. Northern mineral wealth systematically went to underwrite southern development. Locational decisions by the federal government almost invariably favoured the south. Individually these decisions might have been defensible; after all the south had a more highly developed infrastructure, a larger local market and more experienced labour, but the cumulative effect was unacceptable. They were also in themselves adding to a spiral of cumulative causation: the more industry located in Southern Rhodesia the larger its market grew, the larger its work force experienced in industrial employment became, and the better infrastructure became. The economic shortcomings of the Federation cannot all be dismissed as being due exclusively to the political greed of the white settlers of Southern Rhodesia, although that was present. The Federation experienced problems inherent in any attempt to unite territories of greatly differing levels of economic development and, beneath the obvious lack of a political will to achieve a more equitable distribution of resources, there were important problems common to all such economic situations.

In East Africa the British found themselves administering the three large, contiguous territories of Kenya, Tanganyika and Uganda. No attempt was made to establish political integration, perhaps inhibited by the League of Nations mandate status of Tanganyika, but important steps were taken towards economic integration in the form of common public services for all three territories. They included the all-important customs and excise service which, in effect, created a common market and was later extended to a currency board. Further extensions were, for example, to add essential parts of the basic large-scale economic infrastructure, with single common administrations for railways and harbours, posts and telegraphs and later, airways.

The colonial federations did not survive independence. The Federation of Rhodesia and Nyasaland was broken up deliberately to enable the two northern territories to achieve independence. But the break-up also preserved 'responsible self-government' for the white minority in Southern Rhodesia which constitutionally resorted to the *status quo ante*. In November 1965 the Smith regime illegally proclaimed a 'Unilateral Declaration of Independence' (UDI) and, with the support of white South Africa, defied the British government and United Nations' sanctions until 1979. The two French federations (AEF and AOF) did not survive independence and there was no real attempt by either the French or the Africans in the constituent colonies to ensure their survival. The French perhaps had an eye on the benefits of neo-colonialism through dealing with weak individual colonies-cum-states, and the African politicians were concerned to ensure their hold on the colonies, which despite their small size conferred great benefits to the ruling elites, rather than risk all in bidding for the leadership of a larger political entity. In East Africa the common services developed by the British became the basis for the East African Community, set up in 1967 but dissolved in 1977.

Independence came with surprising speed to individual African colonies. The French and British were exhausted by the Second World War, their claims to world leadership finally killed off by the efforts required to win two world wars within thirty years. After 1945 two new super-powers, the United States and the Soviet Union, emerged. Both were overtly anti-colonial and pro-decolonization. The Atlantic Charter, signed during the war, had spelt out the end of colonialism. The tone and sentiments of the Charter were taken up in the Charter of the United Nations, ironically drafted by a Boer, Jan Christiaan Smuts. The British Labour government had no desire to attempt to hold on to India, which was partitioned at independence between India and Pakistan (East and West) in 1947. The French were less ready to relinquish empire and doggedly fought on in Indo-China until the devastating defeat at Dien Bien Phu (1954). Delusions of imperial grandeur again surfaced when the British and the French combined with the Israelis in 1956 to invade Egypt and the Suez Canal in a pathetic attempt to rerun the events of 1882. Nasser was better prepared than Arabi Pasha and the British and French were no longer free agents. The Americans held the trans-Atlantic purse strings and quickly intervened to put the dogs of war on a tight restraining lead. The ignominious retreat from Suez was a significant moment: the British and French were thereafter largely resigned to the loss of empire in Africa. Nevertheless they fought on, particularly in colonies with a large white settler presence. The French gave up Morocco and Tunisia in an attempt to hang on to Algeria. The war in Algeria brought the French government down and General Charles de Gaulle to the Elysée. As is so often the case the right-wing General eventually did the opposite of what was expected of him by bringing Franco-phone black Africa to independence as well as Algeria itself. The British were confronted in Kenya by the 'Mau Mau rebellion' and fought a bitter guerilla war, whilst giving Tanganyika and Uganda independence with equa-nimity. In Rhodesia the British Labour government under Harold Wilson failed to bring the illegal regime of Ian Smith to heel. Wilson, not confident of the backing of his military chiefs of staff, declined the armed option and relied on economic sanctions. That they would be supported by South Africa and Portugal (in Mozambique) was just wishful thinking. Dramatic face-to-face meetings on warships were futile as no agreement could possibly be made with the mendacious Smith.

Elsewhere the French and British went about bringing their black African colonies to independence in contrasting ways. The French colonies were brought to independence individually but more or less simultaneously; fourteen were given independence in the single year of 1960. Independence came irrespective of the preparedness of the individual colonies. Many of the progeny in this multiple birth were premature and weak but the French were eager to nurture them in the swaddling-bands of neo-colonialism.

The British approach was different. Each colony was assessed with care and a bespoke constitution was tailored with enormous attention given to checks and balances of the various political forces within the individual colony. The Westminster model was exported complete with replica mace and other regalia. Much serious effort was made to get the colonies off to a good constitutional start with protracted conferences of all parties to talk through the various options and arrive at a considered settlement. Was the trouble taken by the British over independence constitutions worthwhile? They often did not last. The French model of one constitution for all was at least as effective but perhaps led to blatant neo-colonialism.

The other colonial powers did no better. The Belgian legacy in the Congo was an almost immediate bloodbath; in Rwanda and Burundi the effects of the colonial legacy are all too evident. Portugal and Spain held on to their African territories oblivious of world opinion and the drain of bitter colonial wars. Independence came only after the collapse of their fascist dictatorships in the 1970s. European colonialism came to a messy and protracted end in Africa, delayed a generation until 1994 in South Africa.

# 6

# PROTECTING APARTHEID

During 1960 sixteen African countries celebrated independence, and Macmillan made his famous 'wind of change' speech in Cape Town. During the next three decades South Africa mounted a rearguard action to protect apartheid against the mounting tide of black nationalism. South Africa's action affected all its neighbours: by direct military incursions; by support for civil wars; and by independence delayed, through occupation of Namibia and support for the minority regime in Rhodesia. The period 1960–90 added an unwanted chapter to the colonial experience of southern Africa as the minority government of South Africa sought to protect the apartheid state.

South Africa's relationship with its neighbours, always overpoweringly close, had three historically and geographically distinct phases, which materially affected the colonial experience:

| | |
|---|---|
| 1910–60 | 'acquisitive' |
| 1960–75 | 'windbreak' |
| 1975–90 | 'destabilization' |

Each phase ended in a frenzy of political activity:

| | |
|---|---|
| 1960–64 | Sharpeville, state of emergency, Republic, break-up of Federation, independence for Malawi, Zambia |
| 1975–80 | invasion of Angola, Soweto, state of emergency, independence for Mozambique, Angola, Zimbabwe |
| 1990–94 | unbanning of African National Congress (ANC), release of Mandela, independence for Namibia, negotiations to end apartheid, non-racial elections, majority rule |

In 1910 the British Liberal government gave independence to the Union of South Africa on terms acceptable to and largely suggested by the Boer Generals who had just lost the Anglo-Boer War. Most significant was that the 'colour-blind' Cape franchise was not adopted in the former Boer republics. The Boers had made this a sticking point even before agreeing to sign the Peace of Vereeniging to end the Anglo–Boer War in 1902. The Liberals, many of whom had been against the war in the first place and then felt guilt for its jingoistic excesses, comforted themselves with false hope that in time Boer attitudes would be liberalized.

Britain did withhold from South African control of the High Commission Territories of Basutoland, Bechuanaland and Swaziland at the request of their African chiefs. Provision was, however, made for the: 'possible eventual transfer of the administration of the Territories to the Union of South Africa, subject to certain conditions designed for the protection of native rights and interests' (UK 1952: 5). British government ministers pledged full parliamentary debate and 'that the wishes of the inhabitants would be ascertained and considered before any transfer took place' (UK 1952: 6).

Provision was also made for Rhodesia to join the Union at the end of British South Africa Company rule. In 1910 there seemed little doubt that this right would be exercised and Rhodesian settler representatives attended the constitutional convention of 1909, but a vote on the issue had to be delayed until 1922, when Company rule was due to end. There was, therefore, a clear expectation in 1910 that it was only a matter of time before the Union of the four colonies would be expanded into a Greater South Africa incorporating additionally the three High Commission Territories and at least Southern Rhodesia. When expansion did come it was in a direction not anticipated in 1910.

Despite a pro-German Boer 'rebellion' in 1914 South Africa entered the First World War on the British side. In 1915 Generals Botha and Smuts invaded the contiguous German colony of South West Africa from Walvis Bay, Luderitz Bay and the northern Cape. As the army advanced from the Cape they extended the Cape railway to link with the German colonial railway which was upgraded to the Cape gauge, and became a part of the South African rail network. In December 1920 the League of Nations conferred the mandate of South West Africa on the Government of the Union of South Africa. In Lloyd George's words: 'There is no doubt at all that South West Africa will become an integral part of the Federation of South Africa' (Wellington 1967: 265). At this point there was cause for optimism on the part of those desirous of South African expansion, for the acquisition of South West Africa was an unexpected bonus.

However, the dream of a 'Greater South Africa', which dates from the 1870s, was soon shattered. Hopes of liberalization of the Boers had already been dealt a major blow in 1913 when General Hertzog left the Botha government and founded the (Afrikaner) National Party. Then the violent excesses of the 'moderate' Boer leader General Smuts, in 1921–2 helped swing the largely white settler Rhodesian electorate against joining the Union. Smuts used artillery, tanks and aeroplanes against striking white miners in the Rand Rebellion, caused large loss of life by sending the army in against an African religious sect at Bulhoek and then, in a separate incident, sent bomber aircraft against the Bondelswarts in South West Africa in a dispute which originated in a mass refusal to pay dog licences. Despite a campaign by Smuts to woo the white Rhodesians, they opted instead by 59 to 41 per cent for 'responsible self-government' in 1923. Smuts was of the opinion that the Rhodesian white settlers were: 'afraid of our bilingualism, our nationalism' (Blake 1977: 186). There was probably truth in this perception of his set-back but ironically Smuts during his Rhodesian referendum campaign had been attacked by Hertzog for seeking among the white Rhodesians non-Afrikaner electoral support for the future.

From 1913 South Africa pressed Britain for first the incorporation of Swaziland, which had been a Transvaal protectorate from 1890–9, and later the other High Commision Territories. The pressure grew in the early 1930s when Hertzog was prime minister. The issue became a matter of public debate in Britain where informed opinion swung against incorporation largely because of the perceived Boer attitude towards Africans. The Second World War interrupted the debate, but as late as September 1963 Dr Verwoerd audaciously pressed the point: 'I wish to make an offer to Britain – almost a challenge – to allow us to put to the protectorates what our real policy is and how we view their future' (Verwoerd: 3 September 1963).

The High Commission Territories were vital to the design of 'grand apartheid' as set out in its blueprint, the Tomlinson Commission Report of 1955. Their incorporation, assumed by Tomlinson as foregone conclusion, would have raised the proportion of African land in South Africa from the derisory 13.7 per cent set by the 1936 Land Act to about 45 per cent. Verwoerd maintained that South Africa would: 'lead the territories to independence and economic prosperity far more quickly than Britain could' (Verwoerd: 1963).

But by 1959 Britain, influenced by the accession of Malan's nationalists to power in South Africa in 1948 and the controversial and much-publicized rise of apartheid through the 1950s, had determined that the territories should not be transferred to South Africa. Arrangements were made to bring all three territories to full independence in the late 1960s, ten years before the Transkei, the first 'homeland', was given its unrecognized 'independence'.

South Africa did not achieve her territorial ambitions largely for two related reasons, Afrikaner nationalism and a continued illiberal attitude towards Africans. These antagonized the two constituencies, white Rhodesian voters and successive British governments, who controlled the destinies of the four territories South Africa had been set to inherit. Nevertheless South Africa dominated the British colonial system in southern Africa, largely through the rail network, because although the territories were not incorporated into South Africa all four were land-locked with their traditional means of access to the sea running through South Africa. That condition would have been of no consequence had they become part of South Africa but being land-locked greatly diminished their potential as independent states. Southern Rhodesia sought economic and political development forthright, independent of South Africa. Stiff tariff walls were erected and the Federation of Rhodesia and Nyasaland was, for some, a counter to the emergent apartheid state. Despite this, South African influence strongly permeated not only the three former High Commission Territories but, all of southern Africa.

On 3 February 1960 the British Prime Minister, Harold Macmillan, warned a joint sitting of parliament in Cape Town that 'a wind of change is blowing through this continent'. The speech, interpreted as serving notice on South Africa that her racial policies were not acceptable, heralded a period of trauma both in South Africa and in the rest of Africa. Within South Africa riots at Cato Manor in Durban were followed by disturbances in Pondoland. On 21 March 1960 a peaceful anti-pass laws demonstration at Sharpeville, a black township outside Vereeniging, ended with the police shooting dead 67 Africans and wounding 180 more. The ensuing state of emergency proclamation caused a flight of capital from South Africa. An unswerving Dr Verwoerd, the architect of apartheid who had become South African Prime Minister in 1958, survived an assassination attempt and on 5 October held an all-white referendum on the Republic issue. Narrowly winning (52 to 48 per cent), Verwoerd achieved a long-held Afrikaner dream by making South Africa a republic on 31 May 1961, forfeiting membership of the British Commonwealth on the way. After an initial pounding, international financial confidence in South Africa was restored and investment poured in to launch a period of unprecedented economic prosperity.

Elsewhere in Africa events did not look good for South Africa. The latter half of 1960 saw the newly-independent Congo (Zaire) in a turmoil of bloodshed and convoys of pathetic, mainly Belgian, refugees arrived in South Africa to confirm white prejudices there about black rule. In December 1961 Tanganyika became independent, followed by Uganda in October 1962, and then in December 1963 by Kenya, white settlers, Boers and all. Meanwhile, following General De Gaulle's *volte face* the French settler colony of Algeria became independent in July 1962. At the end of 1963 the settler-led Federation of Rhodesia and Nyasaland broke up, and as a result in 1964 the wind of change reached the Zambesi when Nyasaland became independent as Malawi in July, and Northern Rhodesia became independent as Zambia in October.

A 'windbreak' was erected between black and white Africa. The Caprivi strip, which as part of Namibia was ruled by South Africa, formed part of that frontier, a 55-mile (88km) straight-line boundary with Zambia between the Kwando and Zambesi rivers plus 90 miles (145km) along the median line of the Zambesi. South Africa established a large advanced military base at Katima Mulilo on the Zambesi, opposite Sesheke. The Namibia/Zambia boundary was crossed by no more than dirt tracks. Further east the Zambesi and Lake Kariba formed the 495-mile (792km) boundary

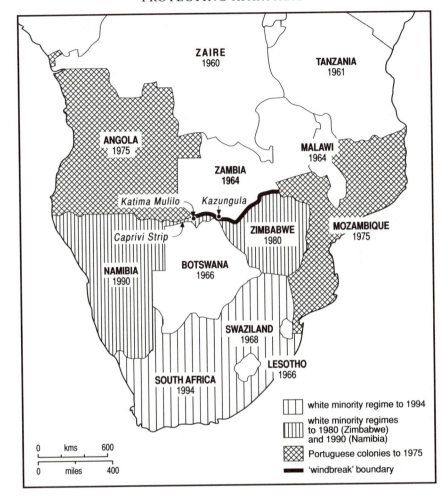

*Map 9* The windbreak frontier

between Zambia and Rhodesia. It was crossed only at three well-guarded places: the rail and road bridge at the Victoria Falls, the bridge on the road between Lusaka and Salisbury (Harare) at Chirundu and the road along the crest of the Kariba Dam.

Along this line the white minority regimes of Rhodesia and South Africa took their stand. Comprising a very large, swift-flowing river, and a vast man-made lake, it was not an easy frontier for guerillas to infiltrate, especially when guarded zealously by the South African and Rhodesian Defence Forces. This 'windbreak' was more than 300 miles (500km) as the crow flies from the nearest point in South Africa, with Rhodesia as a buffer state between giving effective protection to the apartheid state. To east and west, the short windbreak frontier was buttressed by the Portuguese colonies of Mozambique and Angola where liberation guerilla wars raged, but being remote and entirely self-engrossed, these offered no threat to the white south.

South of the Zambesi Britain gave independence to Bechuanaland as Botswana and Bastutoland as Lesotho in 1966 and to Swaziland in 1968. South Africa nullified this by renegotiating the 1910 customs agreement with the three territories. The new *verligte* agreement, which came into effect in 1970, gave South Africa tight economic control over the territories by increasing the revenues which they received under the South African Customs Union effectively to replace British grants-in-aid. Botswana, Lesotho and Swaziland were small and weak, economically, politically and militarily, so whilst their formal independence was an irritation they posed no real threat to South African security. Even so the South Africans were ruthless in demonstrating that they would tolerate no guerilla activity aimed at South Africa from the territories by mounting with impunity 'hot pursuit' across-border raids whenever they deemed it necessary.

That South Africa was nevertheless tense about the 'windbreak' frontier was shown by the Zambesi quadri-point boundary incident of 1970. Where the Chobe tributary met the Zambesi the exact configuration of the international river boundaries was in doubt. Between Zambia and Namibia the Zambesi boundary, established in 1905, was the river thalweg. Similarly the boundary between Botswana and Namibia, agreed in 1890, was the Chobe thalweg. But below the Zambesi/Chobe confluence the boundary between Zambia and Rhodesia, agreed in 1894, was the Zambesi median line. It was not clear whether the median line met one thalweg or the other, or both together at a single tri-point. To complicate matters a fourth boundary, that between Botswana and Rhodesia, approached the Zambesi/Chobe confluence along the line of an old road, the Hunters' Road, and then was extended out into the river in the area of the confluence. The exact line of this last boundary was uncertain because boundary posts along the line of the road as it approached the river had apparently been wrongly placed in 1907. It was a matter of conjecture as to whether the fourth boundary met the median line boundary of the Zambesi below the confluence, or the thalweg boundary of the Chobe above the confluence, or whether all four boundaries lines met to form a single quadri-point. In the first case there would be a common boundary between Botswana and Zambia, in the second there would not, and, in the unlikely event of the last being correct there was a common boundary but it was only a point. Not surprisingly the lines were not accurately mapped, and the largest-scale (1:50,000) modern map available of the area is of no help.

Resolving the ambiguity mattered in practical terms only because the United States planned, in a $13.5 million project due to commence in

1973, to upgrade the road through northern Botswana to the Zambesi at the Chobe confluence where a ferry connected with Kazungula in Zambia. The South African government wrongly thought the plan was to replace the ferry with a bridge. The real intention was to facilitate Botswanan meat exports to Zambia by improving the road and ferry, but South Africa feared that the route would also become a means of escape from, and infiltration to, South Africa which had a long land boundary with Botswana. In a diplomatic note to Botswana, in February 1970, South Africa claimed that there was no common boundary between Botswana and Zambia. Botswana replied firmly, but in low key, that the border and the ferry across it was established 'by long unchallenged usage as well as by law' (Brownlie 1979: 1107). Serious incident was averted, though shots were fired in anger at the ferry. Once again a weak, in this case ambiguous, African boundary briefly became a focus of contention as the political context surrounding it became relevant.

During the period of the 'windbreak' frontier between black and white Africa there were several cross-border clashes, mostly between South African Defence Force personnel (who served in Rhodesia as well as in the Caprivi strip) and guerillas based in Zambia. But for the most part policing of the frontier was effective, so that South Africa materially protected the white Rhodesian regime and, by helping to minimize the external threat to its security, prolonged its life as well as that of the South African regime itself. Thus independence with majority rule for Zimbabwe was delayed by almost sixteen years after that of Malawi and Zambia.

In November 1965 white Rhodesia had made a Unilateral Declaration of Independence (UDI) and was subjected to mandatory international economic sanctions. They were not effective except that they ensured that supplies of oil to Zambia, which had traditionally come from the south via the colonial railway network, most recently from the Umtali (Mutare) refinery, were cut off. Soon after the UDI the Smith regime closed the Zambian border and withheld supplies of coal from the Wankie (Hwange) colliery to the Zambian Copperbelt. This encouraged Zambia to step up the development of its own coal supplies from the Zambesi valley, and to double its efforts to develop a major new route to the sea alternative to those via Rhodesia and South Africa. An oil pipeline was built from Dar es Salaam to the Copperbelt, followed by a tarred road along the same route and then, in 1975, by a railway.

The South African government and the Rhodesian regime did not always see eye-to-eye, especially over Zambia. For example, when Rhodesia cut off coal supplies to Zambia, Dr Verwoerd immediately offered to supply Zambia with coal from South Africa, and so pressured the Rhodesians into changing their minds. The South African government did not want Zambia to escape from its orbit even though it lay to the north of the Zambesi windbreak line. On the other hand the Smith regime took a shorter-term view which, unlike that of South Africa, was conditioned by its contiguity with Zambia. At times this made closure of the common boundary expedient for Rhodesia against South African wishes.

In 1975 the geo-politics of southern Africa was transformed when the two buttresses of the windbreak frontier, Angola and Mozambique, became independent following a military *coup d'état* in Portugal in 1974. South Africa's response to the changed situation created by the Portuguese *coup* was to invade Angola in October 1975 in an attempt to install UNITA (*União Nacional para a Independencia Total de Angola*), one of three rival Angolan liberation organizations, as the official government in Luanda. The South African advance northwards along the coast was stopped by the Marxist MPLA (*Movimento Popular de Libertação de Angola*) assisted by Soviet-supplied and Soviet-advised Cubans. The South Africans were driven back to defensive positions just inside the southern Angola border. Their intervention proved counter-productive as it encouraged Nigeria, followed by the Organisation of African Unity (OAU), to recognize the MPLA as the official government of Angola in November 1975. Angola's independence opened up that country to the guerilla forces of the South West Africa People's Organisation (SWAPO) and presented the South African Defence Force with the task of defending the 855-mile (1,370km) boundary between Angola and Namibia, of which 490 miles (785km) were west of the Okavango river. The front line between black and white Africa was greatly extended and its defence became a major financial drain on South Africa.

In Mozambique there was no rival movement to dispute Marxist FRELIMO (*Frente de Libertação de Moçambique*) becoming the government in June 1975. In March 1976 FRELIMO closed all ports, roads and railways to Rhodesian traffic, including the Malvernia (Chicualcuala) railway direct to Rhodesia from Lourenco Marques (Maputo) along which most of Rhodesia's oil was imported in defiance of the UN sanctions. More importantly, FRELIMO also opened up the 765-mile (1,225km) land frontier of ideal guerilla country between Mozambique and Rhodesia to the armed forces of the Patriotic Front. In anticipation of frontier closure Rhodesia opened a direct railway link to South Africa via Beit Bridge in October 1974 but the days of minority-ruled Rhodesia were numbered. An 'internal settlement' did not gain credibility and the guerilla war progressed. Talks began with Britain, and in December 1979 the UDI ended. Following elections, an independent, majority-ruled Zimbabwe emerged in April 1980, to South African surprised concern under the radical leadership of Robert Mugabe.

Between 1975–80 the frontier of Black Africa moved significantly forward. Mozambique and Zimbabwe both had long land boundaries with South Africa of 305 miles (480km) and 140 miles (225km) respectively, and neither was as easily intimidated as the former High Commission

Territories. Their independence also meant that South Africa's boundaries with Swaziland (267 miles, 425km) and Botswana (1,105 miles, 1,765km) became a little more vulnerable as the source of potential guerilla infiltration was not as easily pinpointed. Only in the west was the front line remote from South Africa itself, and that very remoteness added to the cost of its defence.

South Africa itself experienced major stirrings. In June 1976 the Soweto uprising brought brutal repression to the most widespread black protest witnessed in South Africa. Although the Soweto trigger was the use of Afrikaans as a teaching medium, military defeat of South Africa in Angola and the independence of Mozambique were among other causes. As economic tensions grew the South African government lost its sense of direction and its certainty about apartheid. Various commissions, on labour, trade unions and the constitution desperately tried to reconcile apartheid with a changing economic and political scene. In the midst of this, the 'Muldergate' information scandal led directly to a change of government in South Africa.

Military thinking strongly influenced the new Botha government and radical strategies were unfolded to deal with the potential guerilla threat which might be launched from South Africa's neighbours. The concept of a constellation of states was devised in 1979, envisaging the neighbouring states in a fixed orbit around South Africa, assuming that they would be willing to accept dependency in return for economic help. The South African government set up a Development Bank to channel aid but tied it to the 'homeland' concept. This gave 'independent' homelands the same status as the properly independent states, which was unacceptable to the latter. It also tried to use the aid as a bait to persuade other 'self-governing' homelands to accept 'independence', and this too was rejected.

The constellation concept was positively countered by the front-line states forming the Southern African Development Co-ordination Conference (SADCC), also in 1979. The aims of this organization were deliberately limited: to co-ordinate economic development, independent of South Africa, among the front-line states; and to act as an umbrella for obtaining aid from the industrialized world, notably the Nordic states.

The South African government, realizing that their neighbours were not prepared to accept the dependent role envisaged for them, came to see the apartheid state as under 'total onslaught'. Military thinking concluded it was best combated by 'destabilizing' the neighbouring states through the 'stick' of 'destructive engagement'. In Angola civil war was fermented by direct South African support of UNITA with the backing of the United States. Angola was politically and economically crippled, and forced to commit over 20 per cent of its annual Gross National Product to military spending. From Angolan independence in 1975 the Benguela railway, leading from the Copperbelts of Zambia and Zaire to the sea, has been kept closed. In Mozambique the South Africans took over, from white Rhodesians who gave it birth, the nurturing of the anti-government organization, Mozambique National Resistance (MNR), which waged a devastating civil war. More than one million people became refugees and tens of thousands died of starvation. The effect was to nullify any possibility of Mozambique helping the cause of the ANC in South Africa and, by destroying the Mozambican transport system, to increase the dependence of the land-locked states on South Africa.

Railway lines giving independent access to the sea from the land-locked states were sabotaged to increase the dependence of those states on South Africa. The Nacala railway was closed in 1983 and not reopened to Malawi until November 1989; one span of the long Lower Zambesi bridge at Sena, which carried the railway from Malawi to Beira was blown up in 1986 and has not yet been repaired. The Chicualcuala railway was closed 1984 by guerilla activity. It required a major rehabilitation programme, including a British aid expenditure of £14 million, and was only reopened in January 1991. The Beira corridor from Mutare in Zimbabwe to Beira was kept open by a large Zimbabwean military presence. The port of Beira was sabotaged and operated through the 1980s at reduced capacity.

With the Benguela railway closed, there were two northern rail exits from Zambia and Zimbabwe to the sea, via Zaire and via Dar es Salaam. The former was an inefficient route of rail/river/rail transhipment unable to cope even with Zaire's own copper exports from Shaba, which were exported via South Africa. The route to Dar es Salaam was poorly served despite the railway being newly opened in 1975. Poor maintenance and too few locomotives and rolling stock had reduced the line to fraction of its capacity. Dar es Salaam port was congested and needed major reinvestment. Improvements were slow in coming and Zambia slipped into dependency on South African ports.

South Africa developed other strategies to deal with its immediate neighbours. Swaziland was pressured into signing a secret non-aggression pact in February 1982 with a land deal which would have given Swaziland the South African homeland of KaNgwane and the territory of Ngwavuma which was formally a part of the KwaZulu homeland. The latter, which would have given land-locked Swaziland its own access to the sea, was also intended to be a buffer zone between South Africa and the 50-mile (80km) southern boundary with Mozambique. That ploy became irrelevant and the land deal fell through when Mozambique was persuaded into a similar but much publicized defence agreement, the Nkomati Accord, in March 1984. In both agreements there were clauses to exclude the ANC from the respective countries. The South African aim was to keep the ANC at a distance, and the contiguous countries firmly under control.

Other neighbouring states resisted signing similar accords with South Africa, but were subjected to harassment and heavy pressure by the South

African government. South Africa mounted military raids into Botswana, Lesotho, Swaziland and Zambia, each time killing several people. Known ANC figures and other anti-apartheid supporters were killed by parcel bombs received through the post. Land-locked Lesotho, entirely surrounded by South Africa, was blockaded until a military *coup* ousted its government in January 1986.

The 1980s saw ruthless South African action to minimize the external military threat against the apartheid state. An important, though secondary, aim was to make hostages of the front-line states against the imposition of international economic sanctions against apartheid. The impact of South African destabilization on the front-line states was devastating, setting back development and costing billions of dollars. Not only did the apartheid government extend the colonial period in southern Africa to well beyond the point where most of the states had formally achieved independence but also imposed a regime of destruction on the sub-continent from which it will take a very long time to recover.

Internal unrest and protest, costly external action, international sanctions and disinvestment all took their toll on South Africa's economy. Accommodation between the super-powers made disengagement in Angola possible. As the political and financial costs of maintaining a front-line in southern Angola began to bite, South Africa gave in to international pressure to implement UN Resolution 435 towards Namibian independence. In 1989 free, UN-supervised elections were held in Namibia and independence with a SWAPO government followed in March 1990. Under the agreement the South African Defence Force was withdrawn from Angola and Namibia.

The political pressures from within and without continued to build upon the South African government, Botha's attempted reforms of apartheid made little impact and a state of emergency prevailed from the implementation of the 1985 constitution, which had been designed to introduce some non-whites into central government, but was widely interpreted as keeping blacks out. In 1989 Botha was forced out of office following a stroke and his successor F. W. De Klerk moved quickly to rescue a situation which was getting beyond control. The ANC was unbanned and Nelson Mandela released in February 1990 preparatory to negotiations to end apartheid and minority rule. Most of the legal cornerstones of the apartheid system were quickly demolished but further progress was slow. The process of negotiation was long drawn out, with opposition to a settlement between the two main protagonists, the National Party government and the ANC, coming from the white extreme right and the Zulu-based Inkhata Freedom Party. In a rising tide of bloodshed a date was fixed for universal franchise elections in April 1994. The elections were won by the ANC and in May 1994 an interim government led by Nelson Mandela took office.

The demise of apartheid spells virtually the end of colonialism in Africa. In protecting apartheid from attack from outside the South African government not only prolonged colonialism for its neighbours but, mainly in the 1980s, gave the colonial experience an unparalleled destructive edge. Tens of thousands of lives have been lost in war and war-induced famine, more have been maimed and millions of people have been displaced as refugees. The destruction of the basic infrastructure of the sub-continent has been devastating. With a majority government installed in Pretoria, the civil war in Angola still drags on through several apparent dawnings of peace, and a final settlement in Mozambique is a long time coming. Only when these are finally resolved can the work of reconstruction and rehabilitation really make progress in the sub-continent.

# 7

# THE STATES OF
# MODERN AFRICA

Nowhere in Africa escaped entirely from colonialism during the period 1885–1960. The purist may object to such a sweeping statement, claiming that Liberia has been independent since 1847, but it could be argued that it lost its status as 'a colony for freed American slaves' only in 1980 when the Tubman/Tolbert dynasty, descended from former slave families, was overthrown in a *coup d'état* by the home-grown but ill-fated Master/Sergeant Samuel Doe.

The independence movement in Africa did not get off to an auspicious start when the first post-colonial state to be created was the Union of South Africa in 1910. It was a union of four former British colonies, was very rich in natural resources and had enormous potential for economic development, but was politically flawed in that power was vested almost exclusively, later totally, in a white racial minority. Full independence of the Union of South Africa as a sovereign state came formally in 1929 when the Statute of Westminster defined the position of the white dominions of the British Empire.

Egypt threw off the British yoke in 1922, but Britain retained a military presence in a defined Suez Canal Zone until the 1950s. Ethiopia, having defeated the Italians at Adowa in 1896, retained its independence for another forty years before being forced into the short-lived Italian East African Empire of 1936–41.

Between the two world wars the League of Nations had four African members, Liberia, South Africa, Egypt and (until 1936) Ethiopia. As Ethiopia, apart from the Ogaden region which Britain held on to until 1955, regained its independence in 1941, the Second World War ended with independence in Africa represented by the same four states. That meant that over 88 per cent of the total area of Africa was under colonial rule, over 92 per cent if South Africa was regarded as under a form of colonial rule. In 1994 about 0.05 per cent of Africa was under colonial rule. That included only 14 square miles (36 sq km) of continental Africa in the tiny enclaves of Spanish North Africa. The remainder is in islands, the largest being the Canaries, Madeira, Reunion and Socotra which are

71

claimed to be integral parts of non-African states, respectively Spain, Portugal, France and Yemen.

After the end of the Second World War a climate conducive to decolonization in Africa evolved outside Africa (see Chapter 5). Ideas embryonic in the League of Nations mandate system came to fruition with the setting up of a similar system under the aegis of the newly-formed United Nations (UN). There was no doubt that the UN Trust Territories were to be prepared for full independence and a timetable was laid down that pencilled in 1960 as the year of African independence. What was right for the UN Trust Territories was also appropriate for the other European colonies in Africa. An expectation arose in Africa that independence was within reach.

North Africa led the way when Italy's former colony Libya, after military administration by Britain and France and with UN encouragement, became independent at the end of 1951. Another former Italian colony, Eritrea, was treated very differently. By UN resolution Eritrea was federated with Ethiopia in 1952 and fully incorporated into Ethiopia ten years later. In 1952 King Farouk was overthrown in Egypt. That revolution hastened Britain's withdrawal from the Suez Canal and put pressure on Britain to agree independence for the Anglo–Egyptian condominium of the Sudan which was achieved at the beginning of 1956. The same year saw France giving in to pressure in the Maghreb and granting independence to Morocco and Tunisia. In part this was an attempt to buy time in the settler-colony of Algeria. Thus before the Suez crisis of October 1956, which for many sounded the death-knell of British and French imperialism in Africa, almost all Arab Africa, with the notable exception of Algeria, was independent.

In March 1957 the Gold Coast became the first black African state to attain independence. With geographical licence it changed from its colonial title to the ancient African name of Ghana. Led by one of the great African political leaders, Kwame Nkrumah, the Gold Coast set the pace and pattern of political development in British colonial Africa. It took Nkrumah ten years from returning from self-imposed exile to steer his country to independence. As secretary of the largest independence party he organized demonstrations, was imprisoned by the British, on release formed a new, more radical party and again demonstrated and was imprisoned. His party won the first elections when he was in jail and he was released to head the first internal self-government administration. Nkrumah had to win two more elections before independence when he emerged as Ghana's first Prime Minister. With minor variations this pattern of political progression became the path to independence in most British colonies in Africa. Many other African leaders of stature emerged to meet the challenge of independence: for example, Julius Nyerere in Tanzania (1961), Milton Obote (in his first incarnation) in Uganda (1962), Jomo Kenyatta in Kenya (1963) and Kenneth Kaunda in Zambia (1964). Several literally served time, a jail sentence for political activity almost becoming an obligatory entry in the curriculum vitae of any aspiring African leader. Many of the leaders were of the same generation and had taken over from older men by outbidding them in terms of demands made of the colonial authorities.

The British colonies were treated individually; the timing of independence and the constitution for each new state was different. Special needs were recognized and pains taken to discuss the problems in a relatively leisured way in a manner which fully explored the situation, and then to arrive at balanced solutions aimed at ensuring future political stability.

For example, in Nigeria there was concern about holding the disparate parts of the colony together by devising some form of federal constitution. The Muslim Northern Region came near to going its separate way but in time the desired end was met. The three years that elapsed between the independence of Ghana and that of Nigeria in no way reflected differences in levels of political maturity but rather the greater problems attached to getting the right solution to the constitutional issue. In Uganda it was a matter of balancing south and north, Buganda and the rest, the traditional party led by the Kabaka of Buganda and the modern political movement led by northerner Milton Obote. After elections, at independence, the Kabaka became the titular, non-executive President of Uganda and Obote the executive Prime Minister. In Kenya the problem was perceived as regional and ethno-linguistic: how to counter the political predominance of the Kikuyu. It was achieved by building into the independence constitution regional elements to achieve the balance looked for.

The constitutional results rarely stood the test of time. Within a decade of independence Nigeria was plunged into a bitter civil war, fought on the very issue which had taken so much time in the run-up to independence, keeping the country together. The actual attempted secession came not from the north but from the east, from Biafra. A little over three years into independence Uganda experienced a *coup d'état* when Obote used a then little-known army officer, Idi Amin, to overthrow the independence constitution and drove the Kabaka into an exile which ended (for his son) only in 1993. Obote was in turn ousted by Idi Amin in February 1971. The independence constitution in Kenya was even shorter-lived, lasting for about one year, when the regional safeguards so carefully devised before independence were done away with by the Kenyatta-led Kikuyu-dominated government. The record of the bespoke constitutions for the former British colonies was not very good when judged in terms of their longevity, but it was an approach which at the time seemed to have considerable virtue because it was perceived to promise the emergent states future political stability.

The French approach to African independence contrasted sharply with the British. For the French colonies independence became an issue at the initative of the French government with its *Loi Cadre* of 1956. In 1958 all the French colonies in Africa were asked to vote as individual territories as to whether they wished to be part of the 'French Community', simply *oui* or *non*. The understanding was that if the answer was in the negative then that territory would receive immediate independence. In the event only Guinea had the audacity to say *non*. Guinea was rewarded with immediate independence, but a complete withdrawal of all French assistance, financial aid, administrative personnel, everything. In a fit of Gallic pique it was cut adrift without any preparation whatsoever.

The French-administered UN Trust territories had been scheduled for independence by the UN in 1960, and this timetable was met, with Cameroon and Togo becoming independent in January and April of that year respectively. A further twelve French African territories also were given their independence that year with Senegal, Mali (French Soudan) and Madagascar all becoming independent in June. In near farce, from 1 August an official French government delegation hopped from one colonial capital city to another lowering *le drapeau tricolore* and hoisting independence flags. Dahomey (Benin), Niger, Upper Volta (Burkina Faso) and Ivory Coast, were 'done' at two-day intervals. A four-day rest followed, then another burst of activity with two-day intervals between each celebration of Chad, the Central African Republic, Congo (Brazzaville) and Gabon. Mauritania was sensibly left until the end of November when the weather was cooler.

This approach was independence *en masse*, irrespective of the preparedness or otherwise of the individual colonies. Very often independence was accompanied by a defence agreement between France and the incoming government, sometimes guaranteeing a French military presence which could be called upon in times of need by the new government. This facility was used several times subsequent to independence, the latest in 1994 when French paratroops were flown to Cameroon when a border war with Nigeria threatened. Inevitably there was strong French support, not only in the form of financial aid but also in terms of leaving large numbers of French civil servants in place continuing to run the country, though serving new local political masters. Most of the former French colonies remained members of the franc currency zone, and for these countries France remained very much the main trading partner.

The French approach has been condemned as essentially neo-colonial, with France benefiting from the traditional trading relationship with the African ex-colonies. Raw material imports have been paid for by French exports of manufactured goods. On the other hand, France has proved to be an interested and helpful partner, often ready to intervene, if not always wisely. Successive French Presidents have held annual meetings with the Franco-phone African countries, latterly including those which were not even former French colonies, including Zaire, Rwanda, Burundi, Mauritius and the Seychelles. France used its political strength within the European Community to push through the Lome Convention to the benefit of the trading position of its former African colonies (and France itself). Despite the presence in many former French colonies of French troops and a willingness to use them in support of the government of the day, they have not, on the whole, been any more politically stable than other African countries, with over thirty successful *coups d'état* since 1960. Only Cameroon, Djibouti and Ivory Coast, of former French colonies, have been untouched by this all-too-familiar characteristic of post-independence African politics. On the other hand, economic progress in several former French colonies has been steady, though this is largely due to the fruits of neo-colonialism, sticking to cash crop production rather than venturing to take on the world economic system by bidding for a share of manu-facturing against the might of the industrialized countries. The downside for the former French colonies was illustrated when the French forced a devaluation of the CFA (*Communauté Financière Africaine*) franc by 50 per cent in January 1994 in return for writing off international debts.

The area of Africa, including the offshore islands, is over 11.5 million square miles (30 million square km). Mainland Africa is divided into forty-seven independent states and two other territories, Western Sahara and Spanish North Africa. There are six independent island states and eight other groups of islands off the African coast. Of the area of Africa, 81.7 per cent is divided between twenty-two independent states, the remainder between forty-four different territories. Fifteen independent states together make up an area of only 1 per cent of Africa. In terms of area there are several very large states, the largest being the Sudan, Algeria and Zaire, each of which is greater in area than all twelve countries of the European Union put together. There are also a great many very small states, with seven independent African states actually smaller in area than Wales.

As a continent Africa is sparsely populated relative to, for example, Asia or Europe, with a total population somewhat in excess of 550 million at an average density of about fifty persons per square mile (twenty per square km). Although the continent has been experiencing something of a population explosion since independence, with population growth rates in some countries exceeding 4 per cent per annum, only one African state, Nigeria, has a population in excess of 100 million and only one other, Egypt, a population of over 50 million. On the other hand, seven independent African states have a population of less than 1 million, and another nine have a population smaller than that of Wales. The Seychelles, which is an independent state, has a total population of about only 70,000, roughly equivalent to the population of Hove.

A third measure of the size of states is that of their Gross National Product (GNP). The overall hallmark of Africa here is one of poverty, and the GNP of many African states has actually fallen over the last decade or more, thus widening the gap in levels of wealth between Africa and the industrialized countries. In 1993 at least ten African countries had a total GNP of below US$1 billion. As a basis of comparison the GNP of the United Kingdom was estimated by the same source (the World

Bank) for the same year as US$10 billion, whilst the total GNP for all the countries of Africa together was about one-sixteenth of that of the United States alone. Even allowing for difficulties in calculating GNP in mainly agricultural countries, where a high proportion of people are rural dwellers engaged in subsistence activities, the overall poverty of modern Africa is obvious.

Africa is fragmented into fifty-three separate independent units and the combination of small area, small population and poverty brings with it for many states doubts as to basic political and economic viability. The prospects for independent economic development are extremely poor for the very small economies encountered in Africa. Actual smallness is compounded by the fact that the economies comprise large rural sectors. Many African economies are smaller than the minimum threshold size for the establishment of any manufacturing industry other than small craft or repair and service industries, so a whole avenue of potential development is closed off. Such economies are too small to support even basic services. Independent states such as the Seychelles and the Gambia have difficulty in simply producing sufficiently trained people actually to run the apparatus of state. They cannot aspire to the establishment of a university and any form of higher education has to be very limited. They are almost entirely dependent on other countries for any applied research into areas affecting the basic livelihoods of the majority of their people in agriculture and fisheries. In the event of mineral resources such as oil being discovered the governments of such small countries are particularly vulnerable to the activities of large multi-national corporations whose resources, let alone expertise, dwarf those of the states too small to produce their own experts and often too poor to hire any. This underlines the small and poor African states' dependency, from which there does not seem to be any easy escape.

Large states have their problems too. States the size of the Sudan, Algeria and Zaire cover vast areas and have enormous internal distances. At the time of the Russian nuclear disaster it was pointed out that a straight line the equivalent in length of the distance between Chernobyl and London may be drawn entirely within the Sudan. Some concept of the vastness of the Algerian Sahara was brought home to the British when Mrs Thatcher's son lost himself during the Algiers to Dakar motor rally. The cost of putting in place adequate infrastructure over great distances is considerable and adds to the burdens of poor countries. When size is combined with sparseness of population and resources then the problem is compounded. The cost of long lines of communication has to be borne by few people and by the limited wealth derived from few, dispersed resources.

The problem is not limited to the very largest states but is also the experience of such states as Namibia, which is larger in area than France and the United Kingdom combined and yet has a population of 1.6 million, little over 1 per cent of the combined populations of France and Britain.

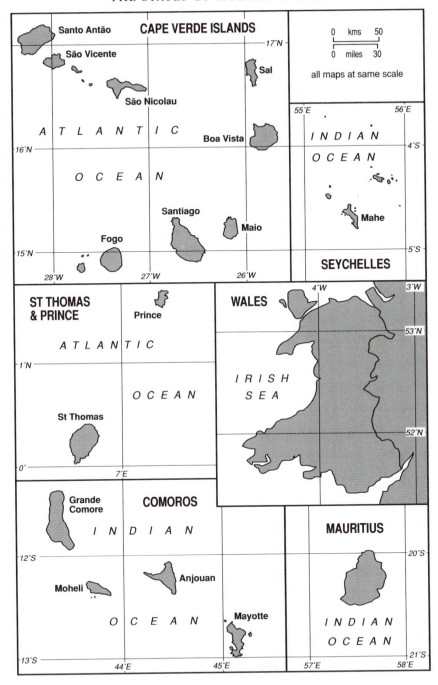

*Map 10* African island mini-states, with Wales for comparison

77

Densities of population for many African states are extremely low, fifteen independent states having population densities of less than twenty-five per square mile (ten per square km), whilst Botswana, Gabon, Libya, Mauritania and Namibia have less than ten persons per square mile (four per square km). Density of population is a measure which is frequently used, but density of wealth, GNP per given area, is not. Nevertheless it is interesting to note that in many African countries the limited wealth is spread very thinly indeed. The extreme cases are Chad and Mauritania, which generated a GNP of less than US$1,500 per square mile (US$600 per square km). Such figures, relating total national wealth and territorial area, give some idea of the problems faced by countries with the combination of poverty, sparseness of population and resources and areal vastness.

Great size can also mean significant disparities between regions within countries. Colonial boundaries often brought together within a single state quite different regions and peoples. The French had the concept of *Le Tchad Utile* for the southern third or so of Chad. It is wetter and capable of supporting greater densities of agricultural population than the arid north. But it also marks a cultural, ethno-linguistic and religious divide which has made Chad difficult to govern as a single state. The Sudan has been plagued by civil war with few breaks since independence in 1956. Again the divide is north/south, Muslim/non-Muslim, more recently fundamentalist Muslim/non-Muslim. This situation has for many years turned drought-triggered famine into mass starvation because the war prevents taking any effective measures to deal with it. Whilst this is to grossly simplify the Sudan conflict it is not basically inaccurate. The roots of the conflict can be traced back to the drawing of a colonial boundary around disparate regions and peoples. Nigeria is another large state, in term of area and population, which has suffered from the same Muslim/non-Muslim divide contributing to secessionist demands. The secessionist problems of Zaire, arising from ethno-linguistic differences, owed much to the sheer size of the country, which encompasses almost all the Congo basin.

Shape is a concept highly relevant to geography and the configuration of so many African states is such that the topic needs to be examined here because the shape of a state could be a significant part of the colonial inheritance in Africa. Many African states are long, narrow and roughly rectangular in shape, usually with one of the narrow ends as a coastline. The shape arises from the ground rules of the scramble for Africa as defined at the Berlin Conference in 1885, when the European powers first defined their spheres of influence in terms of strips of so many miles of coastline. The rules then allowed them to claim as part of their sphere territory inland from that strip of coastline for about 250 miles (400km) defined by boundaries roughly at right angles to the coast. Because on the West African coast the trading posts of the different European powers jostled for position, the primary spheres of

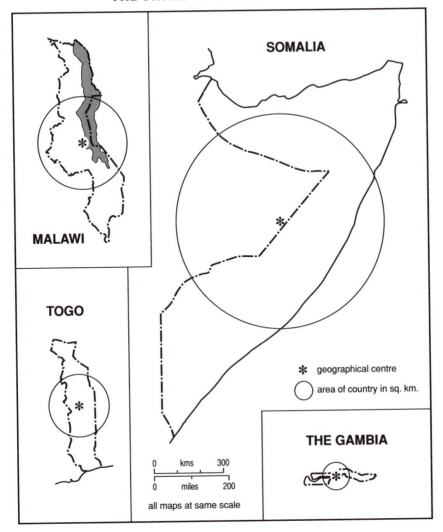

*Map 11* The shape of selected African states

influence were short stretches of coastline which contrasted with the long inland extent of territory claimed.

In West Africa this phenomenon is typified to varying degrees by the independent states of the Gambia, Togo and Benin. The Gambia developed from a British trading post, Bathurst (Banjul), typically located on the small St Mary's Island at the mouth of the Gambia river. To both north and south along the coast French traders had their small trading posts. The Manchester-based traders who had established the Gambia trading post would not be cajoled by the British government into giving up the Gambia in a proposed rationalization of trading posts by which in return for the Gambia the French would have given the British trading posts further east along the West African coast. So when the scramble for Africa rules were agreed the British claimed a sphere based on a very short stretch of coastline on either side of the mouth of the Gambia River. At the coast the British sphere was no more than about 30 miles (48km) wide but inland it became less than half that width, defined as 6.25 miles (10km) from each bank of the Gambia river in a series of arcs of circles. As the inland extent of the sphere of influence was over 200 miles (320km), what emerged as a British colony, and in 1965 as an independent sovereign state, was a long, narrow sinuous enclave surrounded on all inland sides by the former French colony of Senegal. Some Senegalese have characterized the Gambia as 'a sword plunged into the heart of Senegal'. Togo and Benin are less extreme examples but with the same genesis. Togo has a 44-mile (70km) coastline but extends inland for about 340 miles (545km). This ex-German colony partly owes its narrow shape to the fact that it was divided along its vertical axis between the British and the French during the First World War after German colonial forces there had been defeated. The long narrow strip of British Togoland was administered as part of the adjacent British colony of the Gold Coast, and as a result of a UN referendum was incorporated into Ghana. French Togoland was administered separately as a League of Nations mandate, and subsequently a UN Trust Territory. It attained independence as a separate state in 1960. Its neighbour, Benin, the former French colony of Dahomey, has a coastline of about 62 miles (100km) long and an inland extent of about 410 miles (655km).

Land-locked Malawi is also very long and narrow in shape with a length of 550 miles (880km) and a maximum width of 150 miles (240km). That width is where Malawi extends on both sides of Lake Malawi. For the most part the country occupies a much narrower strip of land between the Luangwa watershed and the west bank of the lake only, and south of it the valley of the Shire River. Other African states with strange shapes include Somalia and Mozambique on the east coast of Africa. The shape of Somalia arises from the fact that the independent state comprises two former colonies, British and Italian Somaliland. Both represented spheres of influence the inland extent of which was defined as roughly parallel with the coast, the one on the north coast, the other the east coast of the Horn of Africa. When put together they formed the shape of a figure '7', with head and tail of the numeral about 180 miles (290km) wide. The former Portuguese colony of Mozambique stretches along the east coast of Africa for about 1,450 miles (2,320km) but has a maximum inland extent, along the Zambesi valley, of 460 miles (750km). Mozambique was one of Europe's earliest colonies in Africa and for centuries comprised coastal trading posts and some stations along the lower Zambesi valley as far as Zumbo. During the scramble for Africa the Portuguese administration was limited to the narrow coastal strip by British pressure from Rhodesia and British Central Africa (Nyasaland) but paradoxically was maintained along the coast with British diplomatic help designed to keep out other European powers, especially Germany.

Geometrical expression of areal shape is best done by relating shapes to that of a circle. The shapes of African states are expressed in that form in Table 1. There the measurements are expressed on a scale of 0 to 100, where 0 would equal a perfect circle and 100 a line with length but no width. On this basis the state with a shape least like a circle is the Gambia, followed by Somalia, Malawi, Togo, Benin and Mozambique. States nearest a circle in shape are Sierra Leone, Gabon and Zimbabwe. The question to be answered is whether this is simply a gratuitous exercise in quantification or whether the process of economic and political development in a state is actually affected by the shape of the state.

A major aid to the opening up of a country to economic development is the construction of the basic modern infrastructure of a railway or tarred road. Short feeders of unmade roads quickly and cheaply extend the hinterland, and it is generally reckoned that a major transport artery can positively affect development for at least 50 miles (80km) on either side, plus a radius of 50 miles around the rail/tarred road head. Thus a railway or tarred road 250 miles (400km) long can open up as much as 29,000 square miles (74,000 square km) of territory by accelerating economic development, giving access to markets and facilitating the diffusion of modern farming methods, the use of better seed, fertilizer and so on.

The Gambia, as a colony, was deliberately defined as the territory within 6.25 miles (10km) of the Gambia River. Modern trunk transport in the Gambia is provided by a tarred road from Banjul island, by causeway to the mainland and then roughly parallel with and south of the Gambia River to Basse Santa Su near the extreme east of the country, a distance of about 325 miles (520km). To the south of the road the international boundary with Senegal is never more than about 5 miles (8km) away, whilst to the north the river is on average less than two miles from the road. Because of the position of the boundary alone, the return on the investment in the road in terms of hinterland is reduced to one tenth of what it would have been had there been no political boundary. The nearness of the road to the river in the case of the Gambia reduces the effect of the investment even further. The comparative remoteness of the interior districts makes the task of unifying the state more difficult.

Whilst the Gambia represents the most extreme case, the other elongated states also suffer from the disadvantages of shape. Malawi has been moved to counter this by moving its capital city towards the geographical centre of the state, but Togo and Benin have not. Their northern boundaries are extremely remote from the seats of political power and the centre of

*Table 1* African states: shape, centrality and size indices

| | Shape(s) | Centrality(c) | $\dfrac{S \times C}{100}$ | Size(z) | $\dfrac{S \times C \times Z}{100}$ |
|---|---|---|---|---|---|
| Algeria | 36.1 | 85.2 (5) | 30.8 | 95.1 (2) | 29.3 (2) |
| Angola | 52.2 | 68.0 | 35.5 | 49.8 | 17.7 |
| Benin | 71.1 (5) | 91.8 (1) | 65.3 | 4.5 | 2.9 |
| Botswana | 39.7 | 54.1 | 21.5 | 24.0 | 5.2 |
| Burkina Faso | 54.8 | 6.0 | 3.3 | 10.9 | 0.4 |
| Burundi | 49.3 | 39.4 | 19.4 | 1.1 | 0.2 |
| Cameroon | 60.5 | 29.5 | 17.8 | 19.0 | 3.4 |
| CAR | 61.8 | 40.7 | 25.2 | 24.9 | 6.3 |
| Chad | 48.1 | 57.1 | 27.5 | 51.2 (5) | 14.1 |
| Congo | 64.4 | 61.8 | 39.8 | 13.7 | 5.5 |
| Egypt | 52.5 | 46.9 | 24.6 | 40.0 | 9.8 |
| Ethiopia | 43.4 | 9.8 | 4.3 | 48.8 | 2.1 |
| Gabon | 32.9 | 77.2 | 25.4 | 10.7 | 2.7 |
| Gambia | 86.7 (1) | 71.1 | 61.6 | 0.5 | 0.3 |
| Ghana | 46.4 | 73.2 | 34.0 | 9.5 | 3.2 |
| Guinea | 54.1 | 73.6 | 39.8 | 10.2 | 4.1 |
| Guinea Bissau | 56.3 | 41.7 | 23.5 | 1.4 | 0.3 |
| Ivory Coast | 36.3 | 72.0* | 26.1 | 12.9 | 3.4 |
| Kenya | 47.2 | 38.6 | 18.2 | 23.3 | 4.2 |
| Lesotho | 38.3 | 62.5 | 23.9 | 1.2 | 0.3 |
| Liberia | 52.9 | 55.4 | 29.3 | 4.4 | 1.3 |
| Libya | 44.4 | 77.3 | 34.3 | 70.2 (4) | 24.1 (4) |
| Madagascar | 69.1 | 11.0 | 7.6 | 27.4 | 2.1 |
| Malawi | 80.0 (3) | 53.4* | 42.7 | 4.7 | 2.0 |
| Mali | 53.1 | 69.1 | 36.7 | 49.5 | 18.2 (6) |
| Mauritania | 48.2 | 75.9 | 36.6 | 41.1 | 15.0 |
| Morocco | 68.7 | 29.2 | 20.1 | 17.8 | 3.6 |
| Mozambique | 70.5 (6) | 90.4 (2) | 63.7 | 31.3 | 19.9 (5) |
| Namibia | 41.0 | 4.8 | 2.0 | 32.9 | 0.7 |
| Niger | 68.8 | 85.9 (4) | 59.1 | 50.6 (6) | 29.9 (1) |
| Nigeria | 44.8 | 81.3* | 36.4 | 36.8 | 13.4 |
| Rwanda | 43.8 | 17.1 | 7.5 | 1.1 | 0.1 |
| Senegal | 50.3 | 81.8 (6) | 41.1 | 7.8 | 3.2 |
| Sierra Leone | 29.7 | 79.6 | 23.6 | 2.9 | 0.7 |
| Somalia | 84.3 (2) | 47.0 | 39.6 | 25.5 | 10.1 |
| South Africa | 56.7 | 48.6 | 27.6 | 48.7 | 13.4 |
| Sudan | 35.9 | 26.6 | 9.5 | 100.0 (1) | 9.5 |
| Swaziland | 40.2 | 52.0 | 20.9 | 0.7 | 0.1 |
| Tanzania | 46.7 | 64.3* | 30.0 | 37.7 | 11.3 |
| Togo | 78.9 (4) | 89.5 (3) | 70.6 | 2.2 | 1.6 |
| Tunisia | 66.4 | 69.4 | 46.1 | 6.6 | 3.0 |
| Uganda | 52.7 | 23.8 | 12.5 | 9.4 | 1.2 |
| Western Sahara | 69.9 | 47.7 | 33.3 | 10.6 | 3.5 |
| Zaire | 44.1 | 69.9 | 30.8 | 94.0 (3) | 29.0 (3) |
| Zambia | 45.8 | 30.8 | 14.1 | 30.0 | 4.2 |
| Zimbabwe | 33.9 | 33.1 | 11.2 | 15.6 | 1.7 |

*Notes*: * Calculations for old capital cities
Size: Percentage of area of largest state (Sudan)
Shape: A perfectly circular shape is represented by zero; the less circular the shape of the state the higher the number
Centrality: A capital located at the centre of a state is represented by zero; the more peripheral the capital the higher the number
Bracketed numbers indicate rank

economic growth on the coast. The length and narrowness of the state tends to exaggerate the remoteness of these frontier areas and can lead to feelings of alienation.

Many of the states of modern Africa are disadvantaged by the physical dimensions and shapes which they have inherited. Their physical attributes were quite acceptable when the territories were colonies when the over-arching power of empire was over-riding. But to independent states they become serious impediments to economic growth and unified political development.

# 8

# POLITICAL BOUNDARIES

International boundaries are the most fundamental part of Africa's colonial inheritance. They define the states of modern Africa which, with very few exceptions, are territorially identical to the European colonies they replaced, for all their grotesque shapes and varied sizes. What was acceptable for colonies, which were parts of larger Empires, is inappropriate for independent states. Yet despite the obvious drawbacks of an anachronistic political framework, and the ambiguities, errors and lack of common sense contained in the boundary lines themselves, opportunities for change in the post-independence era have largely been ignored, indeed deliberately thwarted by the African states themselves. In July 1964 the OAU formulated a policy which has prevailed ever since, whereby 'all member states pledge themselves to respect the borders existing on their achievement of national independence' (Brownlie 1979: 11).

Africa has about 50,000 miles (80,000km) of international land boundaries which divide the continent into forty-seven independent states plus Western Sahara and the two tiny enclaves of Spanish North Africa. Thus the boundaries fragment or balkanize Africa into many weak, dependent, political units. There are 106 different land boundaries, three divided into two separate parts. Many of these 109 lines give rise to disputes, bloodshed and even war. They are a constant impediment to good international relations on the continent.

The international boundaries of modern Africa emerged mainly in the thirty years following the Berlin Conference of 1884–5. At Berlin the European powers laid down the rules for their partition of Africa. The boundaries were subsequently drawn by the European powers with scant regard even to the physical geography of Africa, let alone to Africans. In a series of bilateral treaties between the European powers boundaries were drawn to define the different European spheres of influence. From *definition* the process moved through *delimitation*, which drew the boundaries on maps, and finally to *demarcation* (often not completed), which saw them marked on the ground. By 1914 the political map of Africa was virtually complete. The lines still to be drawn were in the Sahara, internal

84

sub-divisions of French colonial Africa and adjustments subsequent to the First World War.

The Saharan boundaries were resolved largely by straight lines. Divisions internal to the French empire came and went, and some were never even drawn, leaving thorny post-independence problems. Lack of precise definition of boundaries was not of major concern when the lines merely divided one part of the French empire in Africa from another. Alignment of some of these 'internal' boundaries depends on administrative practice during the colonial period rather than legal definitions in Acts or Treaties. Colonial maps, which were not always accurate, often afford the best evidence. This lack of precision in the drawing of boundary lines has resulted in several bitter disputes between pairs of former French colonies in north and west Africa, notably Algeria and Morocco, but also Mauritania and Senegal, Burkina Faso and Mali, and Benin and Niger.

The First World War wrought other colonial boundary changes. Jubaland was given by Britain to Italy in 1924 for joining against Germany in 1915. Kenya was compensated in 1926 when its western frontier was moved at the expense of Uganda. France's parallel deal with Italy in 1935, giving part of Chad to Libya, was never ratified, giving rise to a major post-independence dispute, resolved only in 1994. Portugal took advantage of German preoccupation to take back the 215-square-mile (345 square km) Kionga triangle from Tanganyika, seized by Germany only in 1909, literally by gun-boat diplomacy. The territory of Ruanda-Urundi was further taken from Tanganyika and given to Belgium under League of Nations mandate as its reward for being on the 'right' side during the war. France was given the bulk of Kamerun to rule as a League of Nations mandate but also took the opportunity to secure the return of corridors of territory, ceded to Germany only in 1911 to give access to the Congo (Zaire) and Ubangui rivers, to the French Congo. Elsewhere the British Cameroon mandated territory was administered as part of Nigeria and British Togoland as part of the Gold Coast, so changing completely the western boundaries of the two former German colonies of Kamerun and Togo. These changes led to the future boundary problems between Ghana and Togo, and Cameroon and Nigeria.

A few minor boundary adjustments were later made to suit the colonial powers: for example, an exchange between Belgium and Portugal in 1927 of the 'Botte de Diolo' (2,200 square miles, 3,500 square km) to Angola for 2 square miles (3 square km) of the Duizi valley near Matadi to allow a convenient realignment and upgrading of the Matadi to Leopoldville (Kinshasa) railway. Other would-be sensible adjustments were not made because the imperialists could not agree between themselves. Boundary absurdities survive, for example, the Choum railway tunnel in Mauritania which enters and leaves the same side of a mountain range to avoid trespass on or under foreign territory.

Because the boundaries of Africa were drawn with little knowledge of the continent and scant attention to detail, there are many ambiguities in the 50,000 miles of lines. A watershed is a precise enough concept until an attempt is made to delimit it, for example between the Congo and Zambesi basins on a wide, flat, plateau surface. In days before helicopters imperial boundary commissions made heavy work of agreeing boundaries in such terrain, a task made worse where thick vegetational cover blanketed the scene. A river is a river but is the boundary line a thalweg, median line or bank? Which was intended is not always spelt out in boundary agreements, even in formal treaties between two colonial powers. The status of islands (however defined) is surely unambiguous with thalweg or bank boundary, but what if the main channel of the river changes (and it does happen)? Some boundary agreements took this into account, for example, that between Germany (Tanzania) and Portugal (Mozambique). For about 438 miles that boundary is the Rovuma river. The sovereignty over islands in the river was disputed, and in 1913 agreement was reached whereby islands above the Domoni/Rovuma confluence were regarded as German, below the confluence as Portuguese. This agreement was confirmed and clarified in 1936 in an exchange of diplomatic notes between the British (Mandatory Power for the former German colony) and Portugal, which spelt out in great detail the actual position and attempted to account for all eventualities. It defines for the purpose of this agreement a thalweg ('the line of minimum level along the river bed') and an island ('only those which emerge when the river is in full flood and which contain land vegetation and rock or firm soil and are not shifting sandbanks'). Should the river change its course (as defined), it allows 'the Government of the territory prejudiced thereby ... to divert the river into its old bed' or to claim compensation (Brownlie 1979: 978). The Rovuma boundary has not since been a source of dispute. Spelling out the details of the agreements gives some indication as to the complex reality of what might otherwise seem a simple matter – that the boundary is the river! Unfortunately boundary agreements and treaties are rarely spelt out in such detail and innumerable disputes occur because of the omissions. Then some river (or lake) boundaries are defined as one or other bank. Does a river (or lake) bank boundary preclude one country from riparian rights (Namibia where the boundary with South Africa is the north bank of the Orange river), fishing rights (Tanzania where the boundary with Malawi is the eastern bank of Lake Malawi) or navigational rights (Liberia where the right bank of the Cavally river is the boundary with Ivory Coast)? The examples given have all been, or are, matters of dispute between the parties concerned since African independence.

Mountains and other physical features also have to be used carefully as boundaries with precise definition, which in most cases can only be made with reference to the actual terrain. Long-distance definition, as from Berlin,

London or Paris, using seemingly simple geographical terms has caused many problems on African boundaries. The issue was exacerbated by the lack of knowledge of African landscapes and the almost complete absence of large-scale, accurate maps.

If the use of physical features to define boundaries was fraught with difficulties which are still causing problems, it is hardly comforting to know that the other half of Africa's boundaries are comprised of geometric lines: lines of latitude and longitude, other straight lines and arcs of circles. These were employed by the European powers where they admitted they had no geographical knowledge (even a theoretical knowledge along the lines of 'there is a river there that has a thalweg or median line to serve as a boundary'). The more rugged the terrain and the greater the vegetational cover the greater the difficulty of tracing geometrical lines on the ground. Survey techniques were limited, usually deployed by Joint Boundary Commissions, one party (often military) from each of the two European powers concerned. Errors were inevitable, some in basic survey errors, measurement of angles and distance, but also in the precise identification of places where it was agreed measurement should originate. For example, at the far eastern end of the Gambia, where the Anglo-French boundary consisted of arcs of circles, the centre of the circle was placed on the wrong bank of the river. That error was sufficient to prevent the 6.25-mile (10km) arc of circle from intersecting the next meander of the River Gambia to the east and so placing in doubt many square miles of territory, a doubt resolved only in 1976 long after the British and French had departed the scene.

Doubt and ambiguity can easily flare into dispute and hostilities. Ambiguities arise from inaccurate or inadequate surveys but also from failure to ratify agreements. For example, in 1904 the French and the British came to an agreement over the far eastern boundary of the Gambia which gave France many square miles of territory, with the object of giving access to Yarbutenda, the highest point on the River Gambia navigable from the sea. The agreement was never ratified but throughout the rest of the colonial period French maps showed the *nouvelle frontière* despite diplomatic protests from the British and apologies and acknowledgement of the error from the *Quai d'Orsay*. That ambiguity was also cleared up by the successor states (Gambia and Senegal) in 1976. A more fraught example is the Aouzou strip between Chad and Libya which has been a major bone of contention between the two countries from the early 1970s. Here the two colonial powers, France and Italy, came to an agreement in 1935 which shifted the boundary southwards, to give a strip of territory (the Aouzou strip) to Italy (Libya) in return for Italy joining the First World War against Germany. The agreement was never ratified and so did not come into force, but was the basis of Gadafy's claim to the strip in the 1970s which was strongly contested by Chad. War was waged between Libya and Chad intermittently

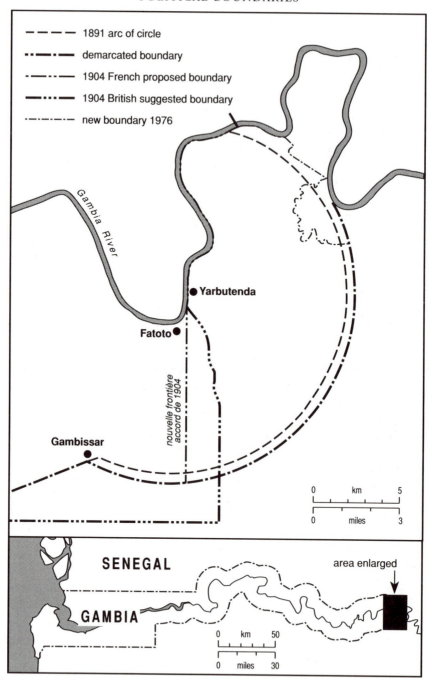

Map 12 The Gambian eastern boundary

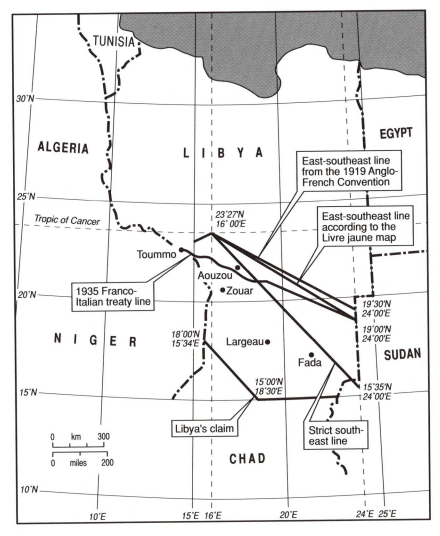

*Map 13* The Chad/Libya boundary

for over twenty years. Eventually the case was brought before the International Court of Justice (ICJ) at The Hague, which in 1994 found in favour of Chad.

Where good relations prevail between neighbouring states minor ambiguities and uncertain boundary alignments are no problem. The border agreement between the Gambia and Senegal in 1976 came at a time when the two states were heading towards federation. The good relations between the two governments, unusually in independent Africa, led to resolution of a potential problem. The converse is also the case: disputes between African states ostensibly arising over boundaries, do not have a boundary as the root cause. When relations are bad between contiguous states a border problem becomes magnified into an issue, a focus for the hostility in reality based on wider disagreement. For example, relationships between Uganda and Tanzania steadily worsened throughout the 1970s from the February 1971 military *coup d'état* in which Idi Amin overthrew Milton Obote. Tanzania gave Obote political asylum and its president Julius Nyerere would have no personal dealings with the Ugandan dictator. Amin for his part resented the fact that Obote was given asylum by Tanzania, accused the Tanzanian government of plotting Obote's return to power and of supporting an *émigré* army ready to invade Uganda. Amin also became very interested in the political geography of Uganda, including its land-lockedness and its boundaries with Kenya and Tanzania. The boundary with Tanzania to the west of Lake Victoria includes one of the many unreasonable situations agreed by the colonial powers, in this case Britain and Germany, and handed down to the successor states, Tanzania and Uganda. The Kagera River, which flows into Lake Victoria, forms the boundary of Tanzania with Rwanda and then Uganda until the river touches the parallel of latitude 1° South. Thence the boundary follows the parallel for 70 miles to Lake Victoria, cutting the Kagera River just before it enters the lake. The river, which is deep and fast-flowing, happens to be a natural divide of culture groups. The boundary arrangement, insisted on by Britain, places a 600-square-mile piece of territory known as the Kagera Salient north of the river, in Tanzania, and a 15-square-mile piece of territory, the Kagera Triangle, south of the river, in Tanzania. The Salient was in 1979 home to about 3,000 people who were ethnic Baganda. Amin aimed to 'liberate' these people from alien (Tanzanian) rule and in November 1978 sent his army to perform the task. It triggered the 1979 war between Tanzania and Uganda which led quickly to the overthrow of Amin. The boundary situation was less than sensible but it had prevailed since the Anglo-German agreement of 1890. Opportunities for realignment had been ignored in the past, in colonial times in 1919 when both Tanganyika and Uganda came under British rule, and at independence for Tanganyika in 1961 and Uganda in 1962. Immediately following independence the thoughts of both governments were on co-operation in the setting up of the East African Community where such frontiers would be unimportant. It was only after the two governments were at odds with each other that the boundary became an issue and provided a focus which led to war. It is likely that even without such an anomolous boundary some pretext would have been found to justify the war but Africa abounds with such 'weak' boundaries which are time-bombs waiting for a change in political circumstance to ignite the fuse. In a change of political context, a point of weakness sometimes becomes significant and a focal point for military

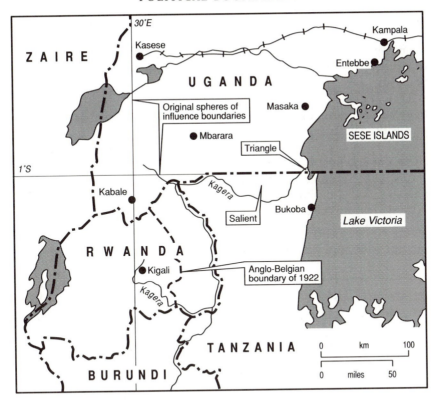

*Map 14* The Uganda/Tanzania boundary

conflict. The mere potential for such disputes weakens African states by apparently legitimizing military demands for scarce resources.

One major effect of the use of physical features (rivers and river basins, 46 per cent and geometric lines, 48 per cent) was to dehumanize the boundaries of Africa. It emphasizes the fact that international boundaries were imposed on Africa from outside by the European powers, and with little knowledge of, or regard to, pre-existing African political and cultural structures and distributions. A checklist shows that the international boundaries of Africa cut through a total of 191 culture group areas, some of which were partitioned by more than one boundary. Put another way, every international boundary in Africa cuts across at least one culture group area, which clearly demonstrates that the phenomenon is continent-wide. To take some examples, no fewer than thirteen culture groups straddle the Cameroon/Nigeria boundary, ten the Kenya/Uganda boundary, while the boundaries of Burkina Faso cut twenty-one culture group areas. The extensive, if sparsely populated, culture group area of the nomadic Berbers is cut by eleven modern international boundaries; that of the more sedentary Swazi, who occupy a much smaller area, is cut by three boundaries. Only rarely do modern international boundary alignments in Africa coincide, for any great distance, with the edge of a culture group or ethno-linguistic area. At the micro-level international boundaries sometimes divide towns, separate towns from their hinterlands, villages from their traditional fields. This despite occasional attempts by boundary commissions (where given the leeway so to do) to make small-scale pragmatic adjustments to a defined and delimited boundary during the process of boundary demarcation. The international boundaries of Africa affect everyday life, particularly in the rural areas.

At a larger scale attempts were occasionally made to redraw boundary lines specifically to coincide with culture divides, for example, the boundary between Tanganyika and Rwanda. Both territories had been part of German

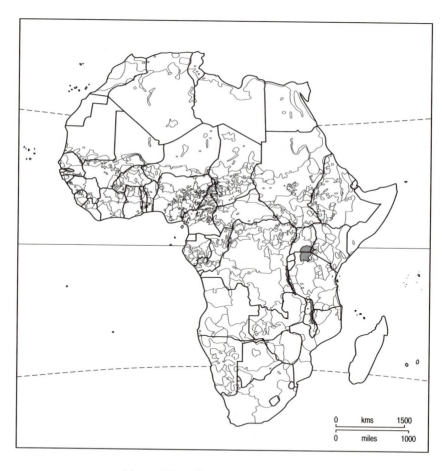

*Map 15* Boundaries and culture groups

East Africa, but were given in 1919 as League of Nations Trust Territories separately to Britain and Belgium respectively. The mandatory powers agreed a boundary line which was approved by the League in 1922 and separate administrations were set up. Almost immediately Belgian missionaries in Rwanda protested to the Permanent Mandates Commission of the League, emphasizing 'the social, political and economic harm [to the Kingdom of Rwanda] caused by the imposition of this arbitrary division', and urged 'the eastward extension of the boundary to the "natural frontier" of the Kagera river' (McEwen 1971: 154–5), which also happened to have been the old German provincial boundary. Belgium and, more reluctantly, Britain, who entertained plans for a railway connecting Uganda and Tanganyika, were persuaded to accept the Kagera river boundary in 1923, although problems of demarcation through the Kagera swamps delayed implementation until 1938. A typically imperialist consideration of a tentative and anachronistic strategic railway was put aside in favour of local human factors because the League of Nations was there to appeal to and was able to ameliorate the harshness of imperialist decisions. Were it that simple! The Kagera River boundary, far from being a natural cultural divide, actually cuts across four small culture group areas.

Given the very large number of culture group areas in Africa (there are more than two thousand different languages) in relation to the number of colonies now transformed into states, it would be impossible to devise boundaries without cutting across culture group areas. To make the task even more difficult, many culture group areas overlap with each other. For example, in West Africa nomadic people such as the Fula or Fulani intermingle with sedentary people like the Mandinka and Wolof in the west and the Hausa in the east. It is common to find villages of two or three different ethno-linguistic groups interspersed. It is claimed that applying linear boundaries to Africa is to introduce an alien concept to people more used to overlap or to unpopulated zones between peoples.

The effect of linear boundaries cutting across culture group areas is deleterious only where central governments interpret the frontiers as impermeable barriers. For the most part African land boundaries are essentially permeable. Modern surface transport, motorable road, rail and waterway, provides Africa with 459 cross-boundary routes, some of which are currently closed. Allowing for some under-counting there is only one cross-border surface route for about every one hundred miles of boundary. Official crossings, that is, those controlled by a customs post or posts, are fewer. Some 345 cross-border roads have customs posts (one to every 145 miles of boundary) so that assuming all rail and waterway cross-border routes are also official, fewer than 400 surface routes are so controlled, one to every 125 miles of boundary. There are other minor (barely) motorable tracks but they are not official crossings and collectively carry very little trade.

Across some land boundaries there is no road, rail or waterway crossing and no official crossing point at all, for example, between the Central African Republic and Congo, Chad and Nigeria, and Tanzania and Zaire. A further eighteen boundaries have only one crossing point each. In contrast six boundaries each had ten crossing points or more. Physical factors strongly influence the frequency of cross-border transport lines. Desert borders, from Algeria's Sahara boundaries with Mali (850 miles/2 crossings) and Mauritania (290/1) to Botswana's Kalahari boundary with Namibia (870/4), have on average as many miles of boundary for every crossing point as did the equatorial forest and large river boundaries in the Congo (Zaire) basin. On open savannah country or high veld the opposite prevails, as between Burkina Faso and Mali (790/13), Niger and Nigeria (955/10), Botswana and South Africa (1200/17), Lesotho and South Africa (555/13) and South Africa and Swaziland (270/11). The high density of official crossings between South Africa and the three former High Commission territories reflects a political policy on the part of South Africa to control tightly those particular boundaries which largely run across open country. They contrast with the Limpopo boundary between South Africa and Zimbabwe which has just one official crossing point but has an electrified fence along the south bank of the large, entrenched but intermittant river.

At the few customs posts officials normally do not stop ordinary people, especially when on foot or on bicycle. All former expatriots will have less than fond memories of hours spent in hot, humid African border posts filling in endless questions on innumerable forms whilst outside hundreds of Africans thronged in both directions without let, hindrance or even a curious glance from the border officials absorbed in stamping foreign passports with great bureaucratic zeal.

People in border areas frequently cross the frontier for a variety of reasons. Where a culture group is divided by an international boundary little attention is paid to the boundary in the course of everyday life. People regularly cross the border, marry spouses and reside across the border for long periods. They attend all manner of ceremonies, social occasions and family celebrations. These activities do not normally concern governments and the cross-border movements involved are seldom hampered. Further, borderland people, whether of the same culture group or not, might cross the border to obtain goods from local stores or co-operatives. Such activity may be incidental to a social visit, or might simply be a matter of distance, for example, the cross-border store being nearer; or ease of access, for example, the store this side of the border requiring a river crossing but not so the cross-border store; or the cross-border store might be larger, offering a wider range of goods, perhaps at lower prices. There are many workaday reasons for such low-level, small-scale, cross-border activity with few deterrents, not different currencies and certainly not language.

These conditions are commonplace in Africa, for example, in the area around the Malawi/Mozambique/Zambia tri-point, where the Chewa and Ngoni people are divided by the boundaries of the three modern states, children cross the border to go to school. Or at the remote eastern end of the Gambia; the furthest east Gambian customs post south of the river is at Fatoto, which is 10 miles (15km) from the easternmost point of the Gambia. The customs office, which is part of the local police station, is set well back from the road. There is no barrier at the post or at the frontier itself, which is demarcated by well-spaced boundary markers of concrete, less than a foot high. They are not visible from the few cross-border tracks (boundaries less recently demarcated are even more difficult to trace on the ground). There are no markings whatsoever on or alongside the tracks where they cross the boundary. Local people know where the boundary markers are and their help is needed to find the markers, even where the boundary line follows the crest of a locally well-defined escarpment. Cross-border traffic by people on foot, bicycle and donkey cart is commonplace. Land is farmed to within a stone's throw of the boundary, which in places is defined in the 1976 Border Treaty in terms of the farmlands of particular villages. Cross-border excursions take place for social reasons (there are Mandinka, Wolof and Fula villages on both sides of the border) and for the purchase of a range of goods at village stores. Such commercial activity is small-scale, with no criminal intent.

Nevertheless 'innocent' cross-border commercial activity can be construed as smuggling under circumstances which independently render the border area politically sensitive. There is a fine dividing line between casual, small-scale movement across a border and systematic illegal trafficking or smuggling. If the price of a commodity is higher on one side of the border, for example because of taxation, then the incentive to indulge in systematic cross-border purchases is increased, as is governmental determination to police the border to prevent the haemorrhaging of significant tax revenue. Similarly if one government imposes a sales tax on, for example, a local cash crop, clandestine cross-border movement of the crop occurs to avoid payment of the tax. Customs officers are deployed in numbers to prevent systematic tax evasion. In such circumstances tension rises in the border area. Customs officials will often be from other parts of the country and ethno-linguistic differences between them and the peoples of the border area can cause heightened tension, with central officialdom at odds with peripheral informality.

All this is short of professional smuggling, where individuals earn a substantial part of their living by illegally transporting goods across a border from one country, to sell at a profit in the other. Where price differentials are great, or there are restrictions on the availability of goods on one side of a border, rewards become sufficient to induce professional smuggling. This in turn leads to the border being 'protected' by customs officers, and affects the borderland people as their own petty cross-border transactions incidentally come under the scrutiny of officialdom. Tension rises throughout the border area as local people get caught up in the situation, many in just going about their normal petty cross-border activities, but some as guides and aides to the professional smugglers and others even as professional smugglers themselves.

It is unlikely that there will be any major redrawing of African boundaries, particularly to take account of divided culture group areas. Contrary to popular belief, few boundary disputes in Africa are caused by divided culture groups. In only a few cases would redrawn boundary lines eliminate ethnic conflict. Most African culture or ethno-linguistic groups cannot be clinically separated by redrawn boundary lines without the type of 'ethnic cleansing' resorted to in Bosnia. The distribution of culture groups is far too complex for any possible simple, simplistic, linear divide. For example, no line can be drawn to separate Hutu from Tutsi in either Burundi or Rwanda, at least before the 1994 genocide in Rwanda, which left many areas 'ethnically cleansed'.

The potent related force of irredentism, where one largely homogeneous ethno-linguistic group seeks to unite all would-be adherents to the group under one political entity is absent as a major cause of dispute in Africa, except in the case of Somalia. In Swaziland and Lesotho, two other states which are largely ethnically homogeneous, with sizeable minorities of the culture group outside the national (nation-state) boundaries, there is no great irredentist cause.

The 50,000 miles of boundary lines in Africa do tend to create a great number of peripheral areas. In these borderlands people, particularly if part of the minority of a divided culture group, can become alienated from political power at the centre of the state. Borderland areas, by definition peripheral, and possibly without any political influence over central resources, can become marginalized economically as well as politically. Borders tend to repel economic development, notably the modern infrastructure of railways and surfaced roads. Borderlands all too easily become economically and politically disadvantaged backwaters simply because they are borderlands.

In Africa many borderlands have been cleared of human population and large areas set aside as National Parks and Game Reserves. A total of twenty-seven different National Parks and Game Reserves are to be found along stretches of eighteen different African international boundaries. For example, the large (7,340 square miles, 19,010 square km) Kruger National Park in South Africa lies along the sub-tropical low-veld boundary with Mozambique, and the *Parcs Nationaux du W* straddle the boundaries near the Benin/Burkina Faso/Niger tri-point in West Africa. The Kruger Park serves as a 38-mile-wide (60km) wide barrier to illegal immigration from Mozambique, the wild animals fulfilling the purpose of the lethal electrified fence erected along the remaining 39 miles (63km) of South Africa's border with Mozambique, which killed an estimated 200 illegal migrants a year between 1986 and 1993 and is still, in 1994, electrified.

Where international boundaries are dividing lines between significant inequalities of wealth, the potential for a large cross-border migration of people is great. The South Africa/Mozambique disparity is by far the greatest (36:1), even allowing that the average figure may be distorted more by high-income whites than by desperately poor homeland blacks. But there are other wide disparities between the GNP per capita of contiguous states in Africa. For example, South Africa is also considerably richer than its other neighbours, Lesotho and Zimbabwe, though only by a factor of four. The Zimbabwe border also warranted a fence, and the South African-engineered *coup* of 1986 in Lesotho indicated the very firm grip the apartheid state kept on that boundary. Tens of thousands of Basotho work legally in South Africa as migrant labour. It is a moot point whether Botswana and Zambia (6:1) have a common boundary. At most it is of only a few hundred yards across the Zambesi where the Kazungula ferry provides the only direct link between the two countries. The wealth disparity between Botswana and Zimbabwe is growing, and is now larger than that between South Africa and Zimbabwe, although job opportunities in Botswana are fewer than in South Africa. There has been a fitful flow of refugees out of Matabeleland into eastern Botswana during the UDI and post-independence periods. Many Egyptians work in Libya officially, but Libya's borders with Egypt (13:1) and Chad (18:1) are protected from illegal influx by the desert, as Gabon's borders with Equatorial Guinea (11:1) are protected by roadless rain-forest.

With the demise of the apartheid regime in South Africa the threat of electrified boundary fences in Africa will, it is hoped, disappear. No civilized state would employ such a lethal device, and few states could commit the resources needed to meet the high cost of other tight boundary controls. Policing illegal immigrants is arguably more effectively and cheaply done by random requests for identity papers inside the country rather than by rigorous checks at the border. The ruthlessness and cost of trying to create impermeable frontiers is beyond the capability of most states and even where they partially succeed clandestine cross-border movements are redirected away from frontal assaults to more subtle means of evasion and entry.

Africa is likely to retain its largely permeable boundaries. The daily cross-border movement of ordinary people is likely to continue without much hindrance. The most effective way to deal with the smuggling of goods and people is a concerted effort on the parts of governments of contiguous states to reduce the disparities which cause the wholesale movement of goods and people. Regional co-operation and setting up common markets is likely to reduce the significance of linear boundaries and enhance border areas. In southern Africa the accession of majority-ruled South Africa to membership of the Southern Africa Development Community (SADC) and the Preferential Trade Area (PTA) will be steps in this direction. Treatment of the causes rather than the symptoms is the direction most likely to succeed, even if it does create wider problems.

That Africa's colonial boundaries have survived over thirty years of independence is due to the OAU accord of July 1964. Whilst it has undoubtedly given a measure of stability to African boundaries it has not prevented border disputes nor frontier wars. On the down side it has also inhibited sensible bi-laterally agreed changes. A new approach might be necessary to encourage measured change before the new climate of political secession and consequent boundary dispute currently prevalent in eastern Europe spreads too far in Africa. Boundary lines and nation-states are alien concepts imposed on Africa. The lines were drawn with a general disregard for local human factors so that disparate culture groups were joined together and individual groups divided, and an incomplete knowledge of the chosen physical features, so that problems and ambiguities, often the cause of strife, were inevitable. Such matters, perhaps the most basically problematical part of the colonial inheritance, can now be best solved in Africa by Africans.

# 9

# CAPITAL CITIES

Capital cities are a large and generally unhelpful part of the African inheritance. Most African capital cities are colonial in origin, which is not surprising given that the states themselves were colonies, but also because of the frequent absence of a pre-colonial urban tradition in Africa. Consequent upon the nature of their origin, African capital cities often occupy sites and locations in relation to the rest of the country which are inappropriate for national capitals, as opposed to colonial capitals. It is another expression of the thought that what best suited a colony does not always adapt well to the needs of a sovereign state, even where that state covers precisely the same geographic territory as the former colony.

Colonial capitals, the seat of alien governments, were often located just within the colony, at the point giving best access to the metropolitan country, for the most part at the nearest feasible coastal site to the metropolitan country. Hence along the West African coast, the eleven seaboard colonies from Senegal to Nigeria had port capital cities. Not only were the colonial capitals on the coast but on long coastlines they were often on that part of the coast nearest the imperial country, for example Lagos at the south-west corner of Nigeria, and Maputo very near the southern extremity of the long, narrow seaboard ex-colony of Mozambique. At independence twenty-three of the thirty-two capital cities of continental seaboard states in Africa (73 per cent) were ports. Only nine capital cities, Asmara, Brazzaville, Cairo, Khartoum, Kinshasa, Nairobi, Pretoria, Windhoek and Yaounde, were non-seaport capitals of emergent continental seaboard states. Cairo, Khartoum and Pretoria have a special reason for being so: they pre-dated European colonial empires. Asmara, Nairobi and Windhoek were colonial capitals chosen for their elevation above sea-level, which made them more acceptable to European settlers.

African capital cities are usually primate cities, very large relative to other cities, large in absolute terms, very fast-growing and faster-growing than other cities. All but five African capital cities are the largest urban centres in their respective countries. The exceptions are Porto Novo (Benin) ranked second, Yaounde (Cameroon), second, Lilongwe (Malawi), second, Rabat

(Morocco), second, and Pretoria (South Africa), fourth. South Africa, a union of four self-governing colonies, at union decided to share its capital city functions between three cities: Pretoria became the administrative capital, Cape Town the legislative capital and Bloemfontein the judicial capital.

The African capital city is often the only urban centre of any note within the state and its population is frequently many times greater than that of the second-ranking city. The following capital cities, for example, are at least ten times larger than the next highest-ranking town in terms of population: Luanda (Angola), Bujumbura (Burundi), Bangui (CAR), Djibouti (Djibouti), Conakry (Guinea), Bissau (Guinea–Bissau), Maseru (Lesotho), Bamako (Mali), Maputo (Mozambique), Kigali (Rwanda), Victoria (Seychelles) and Mogadishu (Somalia). Maputo and Conakry account for over 80 per cent of the urban population of their respective states. For all these countries the rank/size distribution curve is distinctly 'L'-shaped.

Many capital cities are large in absolute terms. Five have a population of over one million: Algiers, Cairo, Addis Ababa, Abidjan and Kinshasa. Only the least populous states have capital cities of less that 50,000 population: Gaborone (Botswana), Praia (Cape Verde), Moroni (Comores), Malabo (Equatorial Guinea), São Tomé (St Thomas and Prince), Victoria (Seychelles) and Mbabane (Swaziland).

Capital cities are almost invariably the fastest-growing cities in African states. Capitals such as Yaounde (Cameroon), Libreville (Gabon), Abidjan (Ivory Coast), Monrovia (Liberia), Niamey (Niger), Lomé (Togo), Kampala (Uganda), Kinshasa (Zaire) and Lusaka (Zambia), all well-established at independence, have increased their population by more than five times since 1960. Even in South Africa, fourth-ranked Pretoria is the fastest-growing city. In five states where the capital is now the largest city, it was not so at the time of independence: Gaborone (Botswana), Nouakchott (Mauritania), Khartoum (Sudan), Ouagadougou (Burkina Faso) and Lusaka (Zambia).

In some cases the high growth rate of capital cities is partly explained by changes affecting statistics, for example, boundary adjustment or more reliable enumeration. But the major part of growth of capital cities derives from the singular attractiveness of these cities as places of paid employment, modernity and, perhaps above all, political power. They often perform many functions: not only capital city, but also port and communications hub and industrial, commercial, educational and cultural centre. Such overwhelming concentration of functions is, in the main, part of the colonial inheritance.

A city that was the colonial seat of government and chief port inevitably accumulated other functions, even where hot humid coastal climates and a high incidence of disease were counterbalancing forces, particularly for Europeans. Army, missionary churches and educational headquarters

100

were all established there, whilst trade and industry, both orientated towards the metropolitan country, rarely had any alternative location. The transformation of a colonial capital to national capital accelerates growth of the capital city as the concentration of functions is reinforced by the addition of the newly acquired instruments and symbols of nationhood and international status. The political and diplomatic function of the capital city in a newly independent state is a vigorous and dynamic force for growth. A modern seat of government immediately surrounds itself with the seemingly endless personnel of legislative and administrative bureaucracy, political parties and diplomatic representation. Secondary and tertiary industries expand at independence, and are free to attract labour with the usual post-independence relaxation of colonially-imposed influx-control. Private and public sector organizations need their headquarters in the capital city for access to politicians and civil servants. Factories and warehouses are located there to take advantage of the modern infrastructure, which is conspicuously absent in most other parts of the ex-colony. Rural–urban migration becomes a direct countryside-to-capital-city movement, especially where the capital city dominates the urban scene.

Two countries in Africa, Botswana and Mauritania, had no colonial capital within their territory. Taking advantage of contiguous, longer-settled parts of the British and French empires respectively, Bechuanaland (Botswana) was ruled from Mafeking (Mafikeng) in the Cape Province of South Africa, and Mauritania from St Louis in Senegal. *New* capitals were created in preparation for independence: Nouakchott (Mauritania) and Gaborone (Botswana). Since independence those cities have forged ahead to become the largest urban centres virtually from scratch, on the basis originally of only the political function, so demonstrating what an engine for urban growth and development the capital city function is. In countries of predominantly rural economic activity, the political/administrative factor is dominant in the process of urbanization. In cities where the political function is one of many, as at Lagos, Maputo, or indeed any of the port capitals, that same dynamic force causes over-concentration of urban growth, which very quickly leads to chronic congestion and inefficiency.

The marginal location of the colonial capital city is not confined to seaboard states. Capital cities of some land-locked states are at or near one frontier and are often main points of entry. Such capitals are at the location most accessible to the former metropolitan power in much the same way as in seaboard states, for example, Bangui (CAR), Maseru (Lesotho) and N'djamena (Chad). Hot, sticky, in the Zambesi valley, but near the point where the railway enters the territory from Cape Town and London, Livingstone was capital of Northern Rhodesia (Zambia) until the higher elevated, cooler, healthier and more central Lusaka was chosen as a new colonial capital in 1931.

The addition of the capital city function to a port or point of entry not only causes over-development and congestion but also concentrates and restricts modern sector development to a remote corner far from the geographical centre of the state. This precisely illustrates the changed perspective from colony to national state. The point most accessible to imperial eyes looking out from Lisbon, London or Paris is often the point least accessible to national eyes seeking the geographical centre of the state into which the colony has transmuted.

Many African capital cities could not be more geographically marginal, more remote from the geographical centre of the state. For example, Porto Novo at the seaboard foot of Benin, Maputo near the southern end of the very long coastal state of Mozambique, Lomé at the corner of the long and narrow rectangular Togo, or Banjul at the sea head of the worm-shaped Gambia. These are states already disadvantaged by their shape. They are doubly disadvantaged by having their port capital city located so peripherally, at a distant corner of the elongated state. Even a land-locked country such as Niger has its capital, Niamey, remote from its geographical centre. Here the disadvantage is distance and peripheral position in an areally large state, rather than location in a disadvantageously shaped state. In Africa few capital cities are located anywhere near the geographical centre of the state. The main exceptions are Windhoek in Namibia, Ouagadougou in Burkina Faso, and Addis Ababa in Ethiopia (see Table 1, p. 82).

Colonial capital cities in Africa were not only marginal in their locations but often occupied sites peculiarly well suited to their original status, for reasons which are no longer relevant. As foreign implants, first as overseas trading outposts and later as the seats of alien government, they needed to be defended against the local people they sought to trade with and then to rule. Island sites were chosen wherever possible to give a measure of natural protection. In the Gambia a small island, St Mary's, at the mouth of the Gambia River was first selected as the site for a trading post/fort in 1662; it became Bathurst. Some 160 miles (255km) upstream another trading post, Georgetown, was developed on the riverine MacCarthey Island (purchased by the British in 1823); the site was chosen with the same defensive considerations in mind. Typically, Bathurst, now Banjul, at the coast, rather than inland Georgetown, became the colonial and later the national capital of the Gambia. This pattern is to be found in many African colony-cum-states. Luanda, Portuguese colonial capital and national capital of Angola, was sited on an island under the protective walls of a massive stone fortress. Mozambique, Portuguese colonial capital of Mozambique from 1505 until 1897, was built on an offshore island with town and anchorage protected by the guns of another large stone-built fortress. Lagos, the British trading post which became colonial and later the national capital of Nigeria, was also built on an island.

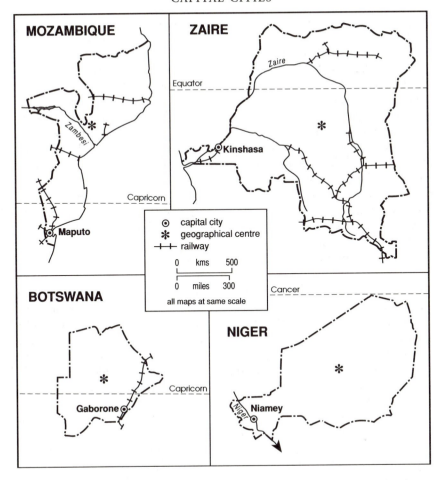

*Map 16* Marginally located capital cities

These off-shore defensive sites served the colonial traders and administrators very well to the end of the nineteenth century. Then, to stimulate and to cope with an increase in trade, new transport technology was introduced from Europe in the form of the railway. Immediately some of the island capital sites became inadequate. Three miles offshore, served by a road causeway submerged at high tide, Mozambique island was perhaps the extreme example. It could not be cheaply or easily reached by rail. Instead a line was started inland from Lumbo on the mainland coast opposite Mozambique to Nampula and Niassa provinces. Eventually a new port was opened up at Nacala some 80 miles (50km) to the north of Mozambique, where the large natural harbour was on the mainland and could be served

103

directly by railway which extended to there. So Mozambique was virtually abandoned by the Portuguese, losing in turn its colonial capital function to Lourenço Marques (Maputo), 1,000 miles (1,600km) nearer Lisbon, in 1897, and its trading hinterland to Nacala, a deep-water port served by rail, from 1950. The situation at other colonial capitals was not so dramatic and not needful of such radical solutions; the railway could be accommodated either by short bridge or causeway onto the island, or by developing the landward part of the harbour. Elsewhere colonial administrations fought to keep their island sites viable by building bridges or causeways to the mainland for rail and road. The bridges for a while liberated congestion of the island sites by allowing urban development on the adjacent mainland. That development often outgrew, especially in areal extent, the old town on the island site. But soon the bridges themselves became congested bottlenecks, slowing traffic to a crawl and imposing restrictions on the development of island-port, colonial-capital sites. The chronic congestion at Lagos, despite post-independence investment in additional bridges and roads and attempts to contain the flow of traffic by a number of drastic measures, is legendary. Many a traveller's tale is told and thought to be exaggerated until confirmed by personal experience. Lagos may be the worst but many other port capital cities are also congested, with difficult access to and from the island core site. The problem is that the nettle has rarely been grasped and that partial, temporary solutions have often been sought: a new bridge here, a causeway there, but always hanging on to the inadequate, restricted site which is patently inappropriate for the varied urban role a capital city is expected to fill in the modern state.

Modern economic development in Africa is usually closely associated with the multi-functional capital city. The process of economic development is not helped by the marginal location of most African capitals and their inadequate, restricted sites. Spread effects have far to travel, the diffusion of modern infrastructure is inhibited, lines of communication are long and tenuous, some regions are incredibly remote. At the port itself there is often insufficient space to meet all the burgeoning demands, especially of modern port developments such as containerization and bulk handling which need very large on-shore storage areas.

Nor is national unity, a prime concern to so many African states, well served by the geographical marginality of the capital city. The ex-colonial capital city tucked into one corner of a new national state is all too easily dominated by people of the culture group of that area. The capital can become regarded not only as remote but alien, representative of one ethno-linguistic group or of a slick isolated urban elite and not in any way symbolizing or helping to create national unity.

The imbalance created by the all-attractive, multi-functional, peripherally located and choked capital city has long been recognized. So has the positive remedial policy option of harnessing the virility of the political capital function to stimulate growth and development in more backward, but geographically more central regions: the Brasilia option. The dynamic capital city force can be relocated in a way port or mineral-based urban complexes never can be, and so can serve as a vehicle for regional development and for better achieving national unity. By removing the most dynamic element in African urban growth the other functions, most of which cannot be moved, are given some space in which to develop. But it is an extremely expensive option that few African countries can afford.

The first new capital cities of post-independence Africa were Gaborone and Nouakchott, already mentioned, hastily built by the departing colonial administrations to cover their embarrassment at having had their colonial capitals outside the territories. They were not, and are not Brasilias in terms of size and scale of investment. More importantly, both occupy geographically marginal locations. Perhaps because of this, because the authorities played safe in terms of location for the new capitals, Gaborone and Nouakchott have been eminently successful as cities, quickly becoming the largest, most dynamic urban centres in their respective countries.

The first African country to embrace the Brasilia option was one of the least likely: land-locked, dependent, poor, conservative Malawi. Lilongwe became the first new post-independence capital city in Africa. The British colonial capital of Nyasaland was Zomba, a typical tropical 'hill station' 2,900ft (884m) above sea level in the hills south of Lake Nyasa. Zomba is about 40 miles (68km) from the main commercial town of Malawi, Blantyre, which is on the railway from the Zambesi and Beira. The decision to relocate the capital was taken in 1965, less than a year after independence, but it was not until 1975 that Lilongwe became the capital in succession to Zomba. Lilongwe is 180 miles (290km) north of Zomba, very near the geographical centre of Malawi, in the widest part of the elongated state and in the home district of President (until 1994) Banda. Part of the planning for Lilongwe was to provide good internal and international communications. There are tarred roads leading north almost to the furthest northern point of Malawi, south to Zomba and Blantyre, to the Mozambique border and on to Tete and Harare, west to the Zambian border and to Lusaka and east to the lakeside which is about 85 miles (135km) distant. The railway has been extended from Salima near the lake via Lilongwe to the Zambian border in anticipation of an eastern railway in Zambia and a new international airport (Kamuzu) has been built to replace Chileka airport near Blantyre. The development of Lilongwe was planned and carried out with substantial financial and technical assistance from a pariah South Africa eager to parade its technical skill and its ingratiating generosity at a time when it was rejected by black Africa as the apartheid state. The price Dr Hastings Kamuzu Banda was prepared to pay was Malawi's diplomatic recognition of South Africa, the only black African state so to do. With a population of about 140,000, Lilongwe is successfully contributing to a geographical redistribution of economic development away from the Blantyre region in the extreme south which had been favoured by the British in colonial times, not least because of its elevated climate.

The nearest Africa has come to a Brasilia in terms of project size is in Abuja, the new capital city of Nigeria. In 1976 Nigeria decided to move its federal capital from the port of Lagos over 400 miles (650km) inland to Abuja, there to create a completely new town in a newly designated

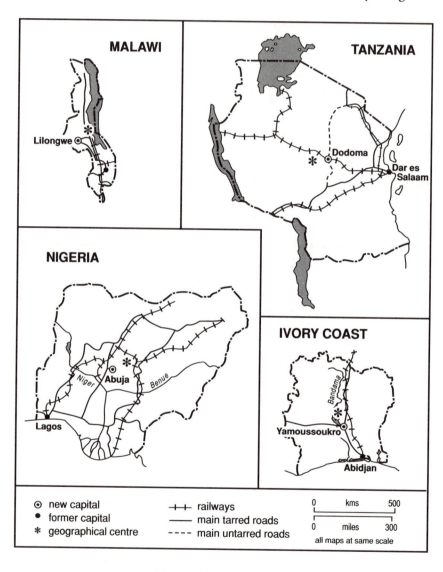

*Map 17* New capital cities

106

Federal Capital Territory. Lagos is a typical colonial capital city centred on an island site, chronically congested, with the city centre severely restricted, and very much on the geographical margin of the large land territory of Nigeria. The move to Abuja was planned to overcome several perceived problems. It will ease the congestion of Lagos and give some breathing space for the industrial and commercial life of the Nigerian chief port to expand. Abuja is located very near the geographical centre of Nigeria and is in the middle belt of the country between the three great regional power blocks of north, east and west. Abuja has a sense of 'neutrality' that Lagos, in the Yoruba heartland of the south-west, never had. Nigerians hope Abuja will become truly the 'symbol of our unity' as the slogan promoting the project claimed. So the new capital city project has both economic and political aims to fulfil.

Abuja is an ambitious and costly project, which includes not only the city itself with lavish accommodation for the large political, administrative and diplomatic functions, but also major improvements in communications. The project was decided upon during Nigeria's oil-boom years and was too advanced to be abandoned when the down-turn in oil prices became evident. Even so the project might have foundered had it not acquired the championship of General Babangida, who was Nigerian military dictator from 1985 to 1993. It was largely Babangida's determination which saw the project through, for he was prepared to make attenuating federal funds available to complete the project; his critics claimed it was to keep rich opportunities for corruption in the form of 'percentage cuts' and 'back-handers' open to his cronies. But Babangida stuck to it and insisted that reluctant politicians, civil servants and diplomats overcame their inevitable inertia, eventually to make the move within a reasonable time-scale. The success of the project is still in the balance but Abuja is now the official capital of Africa's most populous state and there is every chance that it will become a successful capital city. The danger of Abuja becoming an African Brasilia, an expensive flop, is present but rather unlikely. Abuja is not on any Amazonian-like frontier, it is in the centre of things, not only geographically but politically. Whereas Brasilia was a tentative push towards the Amazonian centre of Brazil with nothing beyond, Abuja is between Lagos and Kano, between Kaduna and Enugu, and between Benin and Jos. Its location is well chosen and success should follow.

When Nigeria embarked on the Abuja project it could do so only because its economy was the largest in Africa, because of the boom in oil prices of the early 1970s. The African economy then next in size was that of South Africa, whose government also chose to spend on the creation of new capital cities in the late 1970s and early 1980s. But these were pseudo-capitals for make-believe states, the fantasy 'homelands' of grand apartheid. In creating 'independent' Bantustans, part of the obligatory trappings of imaginary statehood was a 'capital city'. The first 'independent' Bantustan, the Transkei

(1976), had a long-established administrative centre, Umtata. It merely required a new football stadium for the *de rigueur* flag-raising ceremony (at which the flag-pole fell down), a Holiday Inn to house the honoured guests (old National Party hacks), an incongruous tower-block to house the new administration and a brand new University. Bophuthatswana (1977), Venda (1979) and Ciskei (1981) had no equivalent of Umtata and new capital cities sprang up almost overnight from the veld. They were pathetic concrete places attempting to masquerade as symbols of independent political pride and dignity but not succeeding in hiding the reality of the political falsehood and deception of apartheid. In majority-ruled South Africa they have lost even that artificial function and face an uncertain future.

Elsewhere in Africa national budgets do not run to an Abuja or even a Mmabatho (former 'capital' of Boputhatswana). Lack of money accounts for the delay in implementing the decision to move the capital of Tanzania to Dodoma from the old German-selected colonial port/capital of Dar es Salaam. At Dodoma the east–west Tanganyika railway crosses the 'Great North Road'. It is 287 rail miles (460km) west of Dar es Salaam and 267 road miles (427km) south of Arusha, near the geographical centre of the country. The small town of Dodoma is the long-established colonial district and provincial headquarters of a poorly developed region which once acquired fleeting notoriety as the centre of the British Colonial Office's groundnuts scheme fiasco in the late 1940s. Poverty-stricken Tanzania, unlike Nigeria, cannot throw hundreds of millions of pounds at the scheme and it will take some time to develop. Only in the 1990s was the 'Great North Road' between Dodoma and Arusha surfaced for the first time, whilst the southern stretch to Iringa (161 miles, 258km) is murram. To the west the road network remains rather primitive and the airport at Dodoma only caters for internal flights. Communications, Dodoma's main attraction, need great improvement before a new capital city can emerge, in addition to work on Dodoma itself to accomodate the basic political, administrative and diplomatic needs of a new capital city.

A homely variant on the capital city theme is that of Yamoussoukro in Ivory Coast. A small provincial town 166 miles (266km) inland from the port capital of Abidjan, Yamoussoukro commended itself because of its more central location within Ivory Coast, but also because it was the birthplace of the late President Felix Houphouet-Boigny. It is the site of the president's extravagant cathedral, whose dome, by order of Rome, is just smaller than St Peter's. Yamoussoukro is on the northern edge of the rain-forest and the road connection with Abidjan is being upgraded to autoroute standard. Nearer the geographical centre of Ivory Coast, with an international airport, on the railway as well as the main road is Bouake, 66 miles (106km) north of Yamoussoukro. It seemingly lacks only the all-important accident of birthplace.

The inheritance of the colonial capital city has had a deleterious effect on many African states. That few have so far taken positive action to relocate their capital city is principally a matter of cost, particularly when so many African countries are poor and face other, more urgent, demands on scarce resources. Few African countries could find the vast sums necessary for capital relocation, especially with all the related reorientation of national and international communications that are often necessary.

A second factor is that capital city relocation is not always an immediate success or even a success at all. Advantages apparent in theory are not easily demonstrated in practice and are certainly not guaranteed. The uncertain promises of long-term gains pale against the certainty of considerable short-term capital costs. The spectre of Brasilia is well known throughout the world. Placed against other candidate projects in competition for scarce capital resources, capital city relocation is likely to fare badly.

Inertia is another major factor which tends to increase with time from independence. As the existing capital city attracts more investment and growth the more reluctant a government might become to undertake radical change. On the other hand, progressive congestion often makes the problem more acute and might eventually force serious consideration of capital city relocation. Where the colonial capital city happens to be in the heartland of the dominant culture group there is less enthusiasm for change even if the location is far from the geographical centre of the state.

In many states there is no alternative viable location even if funds were available and for others relocation simply would not be appropriate. For example, in Botswana it would not be realistic to relocate the capital city away from the south-eastern line of rail where the bulk of the population lives. The Kalahari desert to the west and the Okavango swamp to the north, whilst they possess some exciting long-term development potential, are not candidates for capital city relocation in the foreseeable future.

Nevertheless capital city relocation remains one of the few options available to governments wishing to redress geographical imbalances in development, at the same time relieving the strangulating congestion of existing capitals. Unlike many other major urban functions the political and administrative function is mobile. That said, it is an extremely costly and uncertain solution. The experience of those making bold to move will be carefully scrutinized but as time goes on the cost of changing location is likely to increase significantly.

# 10

# LAND-LOCKED STATES

Between 1960 and 1980 African independence brought into existence four-teen land-locked states and so doubled the number of such states then in the world. In 1993 the independence of Eritrea made Ethiopia Africa's fifteenth land-locked state.

The land-locked states are peculiarly vulnerable. They are the part of the African inheritance which has caused chronic problems related to the creation of small, weak, non-viable states and some acute problems attaching to individual states. The experiences of African land-locked states pale against those of parts of former Yugoslavia and the Soviet Union in the 1990s but it was in Africa for the first time in the post-Second World War period that the niceties of legal argument about right of access were supplanted by the brutality of *realpolitik*. In 1960 Africa's first modern land-locked state, Mali, immediately had its traditional access route to the sea closed by its seaboard neighbour, Senegal. In January 1986 Lesotho, a state entirely surrounded by South Africa, was successfully blockaded by its much stronger neighbour to effect the overthrow of the Lesotho government. Those two events, which span much of African independence so far, underline the painful travail of the land-locked states.

The land-locked states of Africa are located in three separate blocks of contiguous states: west, east and south central, plus the isolated land-locked states of Ethiopia in the Horn of Africa, and Swaziland and Lesotho in southern Africa.

| West | East | Horn | South central | South |
|------|------|------|---------------|-------|
| Mali | Uganda | Ethiopia | Malawi | Swaziland |
| Burkina Faso | Rwanda | | Zambia | Lesotho |
| Niger | Burundi | | Zimbabwe | |
| Chad | | | Botswana | |
| CAR | | | | |

The western block of five land-locked states was in colonial times all French. Mali (French Sudan), Burkina Faso (Upper Volta) and Niger were part of the *Afrique Occidentale Française* (AOF) and together cover a vast, sparsely populated area between the coastal states of West Africa and the Sahara. Chad and the Central African Republic (CAR) (Obangui Chari) were part of the *Afrique Equatoriale Française* (AEF). Chad is also vast, covering large parts of the central Sahara as well as some savannah country in the south. The three western territories resulted from the French imperial thrust from Senegal to and along the inland Niger, aimed deliberately at preventing inland expansion by other colonial powers from the coast. Large parts of the inland territories were for many years not given the status of colonies but were under French military rule. Upper Volta appeared (1919), disappeared (1932) and then reappeared (1947) on the political map at the changing whim of French administrators. Thinly populated and sparsely ruled, the territories were remote and possessed too few natural resources to attract much colonial attention. Economic development was very limited and the colonial adminstrations made few improvements. The land-locked territories were penetrated by two railways from the coast, from Dakar in Senegal to Koulikoro on the navigable upper Niger river in French Sudan, and from Abidjan in Ivory Coast to Ouagadougou in Upper Volta. A second French railway to the Niger from Conakry in Guinea did not enter French Sudan. Obangui Chari (CAR) and Chad were even more remote and less well-endowed than their west African counterparts. All were colonial culs-de-sac, extremities of far-flung empire. Independence came to the individual colonies, converting land-locked colonies into land-locked neo-colonial backwater states not even parts of an over-arching empire.

In East Africa Uganda was a British protectorate where Captain Lugard, indirect rule and Christian missionaries flourished. Its protectorate status meant there were very few white settlers. Uganda possessed few mineral resources but the southern part of the protectorate, the kingdom of Buganda, was agriculturally very rich. Cash cropping of cotton, coffee, tea and sugar were encouraged by the colonial administration and to some extent by the Uganda railway built from Mombasa, at first to Port Florence (Kisumu) on Lake Victoria, but by 1935 extended to Kampala. Rwanda and Burundi result from the division of the League of Nations Trust Territory of Ruanda–Urundi which in turn had been carved out of German East Africa (Tanganyika) and given to Belgium after the First World War. Ruanda–Urundi was similar in several respects to Uganda. There were few settlers but many Christian missionaries; few mineral resources but rich agricultural land and a high density of population. It was very remote with few development prospects. Belgium made few infrastructural investments. At independence little changed except that one mandated territory was divided into two small states, both land-locked.

Ethiopia was land-locked but independent after the scramble for Africa. It gained access to the sea by absorbing the former Italian colony of Eritrea with the connivance of the United States, which specifically argued for Ethiopia's right of access to the sea. The Eritreans never accepted their fate and fought long and hard for their independence. When that was achieved against all odds in 1993, Ethiopia reverted to its former status as a land-locked state.

Nyasaland (Malawi), Northern Rhodesia (Zambia) and Southern Rhodesia (Zimbabwe) comprised the short-lived British Federation of Rhodesia and Nyasaland (1953–63). In 1964 Malawi and Zambia attained independence separately, leaving a renamed Rhodesia, replete with large white settler population, to follow a troubled fifteen-year path to independence with majority rule as Zimbabwe. Nyasaland was a British protectorate, with few settlers but many missionaries. Like Uganda, Rwanda and Burundi it possessed few mineral resources but a fairly rich agriculture with important cash crops, notably tea and tobacco, and supported a high density of population. Access was via the Zambesi and Shire valleys and, after the completion of the Lower Zambesi Bridge in 1935, from the Portuguese port of Beira. Northern Rhodesia was also a British protectorate but was athwart the railway to the Katangan copper resources. Northern Rhodesia became a major mineral producer itself when the Copperbelt was opened up from the late 1920s. European farmers settled along the line of rail, and the copper companies developed mines and mining camps which mushroomed into large towns, to give some of Northern Rhodesia at least a character different from that found in other British protectorates in Africa. The modern sector was not sufficient to delay independence, but its specific form of development was to create problems for an independent land-locked Zambia. Access was mainly from the south via the Cape railway, although alternative (west coast) routes had been developed during the inter-war period through the Belgian Congo (Zaire) and Angola. Southern Rhodesia was white settler country where half the land area was alienated. Early hopes of great mineral riches were never fully realized and settlers were given land in compensation. Nevertheless Southern Rhodesia developed a significant mining sector, as well as a rich, settler-based commercial agriculture and a well-developed manu-facturing sector. Rhodes fought to prevent the Rhodesias being land-locked, but all his efforts to buy, bluff or force a corridor of access to the coast through Mozambique came to naught. So Southern Rhodesia remained land-locked and, in terms of its level of economic development and settler population, was anomolous among Africa's land-locked territories. On the other hand, it was long seen as a potential fifth province of South Africa and as such would have been similar to the other land-locked provinces of the Transvaal and Orange Free State. But in 1922 the settlers rejected union.

Bechuanaland (Botswana), Basutoland (Lesotho) and Swaziland also escaped union with South Africa. Bechuanaland protectorate was of vast inland extent, covering much of the Kalahari desert. Its very sparse population lives mainly near its south-eastern border. Soon after independence Botswana was economically transformed, through mineral discoveries, from colonial missionary backwater to the richest land-locked state in Africa in terms of

GNP per capita. Rhodes' rail 'road to the north' provides basic infrastructure along the eastern strip of Botswana and gives vital accessibility from the south.

Swaziland is a small independent state loosely based on the traditional Swazi kingdom. It became a British protectorate following the Anglo-Boer war after a period in the 1890s when it was administered by, but not absorbed into, the South African Republic (Transvaal). Before independence more than half the land in Swaziland was alienated, largely by white South Africans. In the 1930s asbestos deposits were developed but did not open up Swaziland partly because they were marginally located and exported by aerial ropeway across the border to the South African railhead at Barberton. Iron ore mining, beginning near the end of the colonial period in the 1960s, did bring with it a direct rail connection to the port of Lourenço Marques (Maputo) in Mozambique. Basutoland (Lesotho), another African kingdom which emerged during the early nineteenth century *Mfecane*, sought British protection against Boer encroachment from the Orange Free State in 1868. Lesotho has few natural resources and survives economically mainly by exporting labour to the mines of South Africa. A rail spur from the South African system extending one mile into Lesotho facilitates this human traffic. Among the land-locked states of Africa Lesotho is unique and doubly disadvantaged as it is completely surrounded by a single country, South Africa.

The land-locked states of Africa, with few exceptions, may be characterized as small (not necessarily in area), economically and politically weak states, generally under-developed and dependent. Eleven land-locked states are among the poorer half of African states in terms of GNP per capita. The only land-locked state among the ten richest African states by this measure is Botswana (fourth). Only Zimbabwe and Ethiopia among the land-locked states had a total GNP in excess of US$5,000 million in 1993, and so ranked respectively as the twelfth and fourteenth largest economies in Africa. By another measure of size, Ethiopia has a population of almost 50 million whilst Uganda alone of the other land-locked states has a population of over 15 million. The land-locked states also seemed to lack political influence. On average, land-locked states had less than half the number of resident diplomatic missions from other African states than the seaboard states of Africa, despite having on average more direct neighbours or a higher contiguity factor. By most measures of size, wealth and influence, the land-locked states of Africa stand out as being among the weaker states of the continent.

Thus in addition to being dependent on other states for access to the sea, the land-locked states are usually weaker than their seaboard neighbours in most respects. This makes them much more vulnerable to pressure from those neighbours. The David and Goliath relationship between Lesotho and South Africa is an obvious extreme but other ill-balanced relationships between land-locked states and their access neighbours are to be found

*Table 2* Land-locked states: population, size of economy, per capita
GNP and resident African embassies

| Land-locked state | Population (millions) | Total GNP (million US$) | GNP per capita (US$) | Contiguous states | Resident African embassies |
|---|---|---|---|---|---|
| Mali | 9.2 | 2,744 | 300 | 7 | 9 |
| Burkina Faso | 9.8 | 2,928 | 300 | 6 | 5 |
| Niger | 8.4 | 2,313 | 270 | 7 | 8 |
| Chad | 6.1 | 1,248 | 200 | 6 | 7 |
| CAR | 3.2 | 1,263 | 390 | 5 | 9 |
| Uganda | 18.0 | 3,486 | 190 | 5 | 10 |
| Rwanda | 7.5 | 1,499 | 200 | 4 | 8 |
| Burundi | 6.0 | 1,102 | 180 | 3 | 4 |
| Ethiopia | 53.3 | 6,144 | 100 | 5 | |
| Malawi | 9.3 | 2,034 | 220 | 3 | 5 |
| Zambia | 8.5 | 3,152 | 370 | 7 | 13 |
| Zimbabwe | 10.6 | 5,756 | 540 | 4 | 17 |
| Botswana | 1.4 | 3,630 | 2,590 | 4 | 5 |
| Lesotho | 1.9 | 1,254 | 660 | 1 | 0 |
| Swaziland | 0.9 | 933 | 1,050 | 2 | 1 |

*Source:* World Bank Atlas, 1995

throughout the continent, for example, between Botswana and South Africa, Uganda and Kenya, Chad and Cameroon or Nigeria, Burkina Faso and Ivory Coast, and Mali and Senegal or Ivory Coast.

The key characteristic of land-locked states is, by definition, lack of direct access to the sea and therefore dependence on another state to grant access over their territory. Even so some land-locked states are more remote than others and some have alternative routes of access whilst others do not. Chad and Rwanda are more than 1,000 miles (1,600km) from the sea by any overland route. Botswana, Burkina Faso, Burundi, Mali, Niger, Uganda and Zambia are all more than 500 miles (800km) from the sea. On the other hand the Swaziland capital Mbabane is only about 225 miles (360km) from the Mozambique port/capital of Maputo. Most remote in terms of quality of route are Burundi and the CAR, because their only means of access other than by minor murram roads is by lake or river transport from distant railheads in other countries. What is more, like Lesotho and Uganda, they do not have viable alternative routes through other access states.

Some land-locked states in Africa suffer from the fact that their traditional colonial access route was longer and involved more trans-shipments from one mode of transport to another than the most direct route through the colony of another power. This was well illustrated quite early in the colonial period in West Africa when during the First World War the French faced a major rebellion in Niger. Troops were sent from Senegal along the traditional,

114

*Table 3* Selected land-locked states and their seaboard neighbours

| Land-locked state | Seaboard neighbour | Population | GNP | Rank among African states GNP per capita | Embassies |
|---|---|---|---|---|---|
| Lesotho | (South Africa) | 39 (4) | 37 (1) | 19 (5) | 51 (48) |
| Botswana | (South Africa) | 41 (4) | 18 (1) | 6 (5) | 37 (48) |
| Uganda | (Kenya) | 11 (10) | 19 (10) | 43 (36) | 22 (7) |
| Chad | (Cameroon) | 29 (16) | 38 (8) | 41 (15) | 33 (12) |
| Chad | (Nigeria) | 29 (1) | 38 (4) | 41 (35) | 33 (2) |
| Burkina Faso | (Ivory Coast) | 19 (14) | 24 (9) | 34 (19) | 37 (6) |
| Mali | (Senegal) | 21 (26) | 25 (11) | 35 (16) | 26 (9) |
| Mali | (Ivory Coast) | 21 (14) | 25 (9) | 35 (19) | 26 (6) |

*Source*: World Bank Atlas, 1995

but slow, French colonial route from Dakar via Bamako and the Niger river but an emergency advance party was rushed in arrangement with Britain (then a wartime ally rather than a colonial rival) by sea to Lagos, thence by rail to Kano and so on to Niger.

Throughout the colonial period Chad was normally reached by the all-French route of rail from Pointe Noire on the Gabon coast to Brazzaville in the French Congo (Congo), thence by river boat to Bangui in Obangui Chari (CAR) and finally by unsurfaced road to Fort Lamy (N'djamena). It passed through four colonial territories (all now independent sovereign states), with two trans-shipments over a total distance from the sea of about 1,810 miles (2,900km). A shorter alternative rail/road route through Cameroon was developed after the former German colony had become a French-administered League of Nations mandate. The most direct route from the sea to Chad, through Nigeria, is about 1,085 miles (1,735km) by road with no trans-shipment necessary, or just one if the goods are sent by rail to Maiduguri. The all-British colonial route from the Cape to the Copperbelt by rail was about 2,150 miles (3,440km) compared with the post-independence Tanzam route from Dar es Salaam of about 1,130 miles (1,810km). The former Belgian-administered Mandated Territories of Rwanda and Burundi are twice as far from Matadi on the west coast through former Belgian territory than from Dar es Salaam or Mombasa on the east coast.

Since independence several land-locked countries in Africa have found it necessary, or at least highly desirable, to develop alternative shorter routes to the coast. They include the Tanzam route from Zambia on which pipeline, tarred road and railway were completed between 1968–75, and the new rail link between Malawi and the Nacala line completed in 1970. Roads from Mali to the Abidjan railway, from Rwanda into southern Uganda, and from Burkina Faso into northern Ghana have all been upgraded. The construction and improvement of new and alternative routes has imposed a considerable financial burden on what are essentially poor states. That part of the colonial inheritance which has decreed their land-lockedness has placed these states at a considerable disadvantage.

Accessibility is more than the *existence* of a route, even where the route is over the territory of a friendly neighbour. Bottlenecks arise at sea ports, river ports and river ferries, beyond the control of the land-locked state. Unintentional delays are caused at customs and immigration posts by the welter of bureaucracy in which such places delight. Customs duties, freight rates, road transport licensing, railway rolling-stock deployment and operating schedules require bilateral agreement and implementation. Maintenence on vehicles, locomotives, signalling and the route itself needs to be systematic. All these operational matters, let alone areas of capital investment and route development, call for the closest international co-operation and co-ordination. This is rarely achieved, sometimes just because of inefficiency, but also because access states often differ from their land-locked neighbours over priorities for the allocation of scarce resources. All such matters are potential sources of international friction and, because their vital external trade arteries are at risk, are most serious for the land-locked states.

In 1959, during the period of internal self-government between French rule and independence, the two former colonies of Senegal and French Soudan united to form the Federation of Mali, and as such achieved independence on 20 June 1960. Exactly two months later the Federation broke up and the two states went their separate ways as Senegal and Mali, the latter becoming on 22 September 1960 Africa's first post-colonial land-locked state. The bitterness arising from the split was such that Senegal closed Mali's traditional colonial route of access to the sea, the Dakar to Koulikoro railway which had been completed from the Atlantic coast to the navigable upper Niger in 1914. Before 1960 about 80 per cent of Mali's external trade, mainly groundnut exports, had passed along this route. The trade had to be rerouted. Two possibilities were considered: via the Guinea railhead higher up the navigable Niger at Kouroussa, and via Ivory Coast by means of road transport to Ouagolodougou, which is on the Abidjan railway. The latter was chosen despite its greater length because the Guinea railway was not able to handle the bulk of Mali traffic. The 355-mile (570km) route from Bamako to Ouagolodougou was upgraded to a tarred road and Mali invested in a fleet of 400 lorries to handle the trade. The Abidjan railway encouraged the additional traffic by offering a preferential tariff to offset any advantage the Guinea route might have acquired with improvements. In 1961 the short-lived union between Ghana, Guinea and Mali led to serious consideration of improving the Guinea railway and extending it to Bamako. Soviet engineers surveyed the route but the project was never carried out, not least because the Abidjan route was working

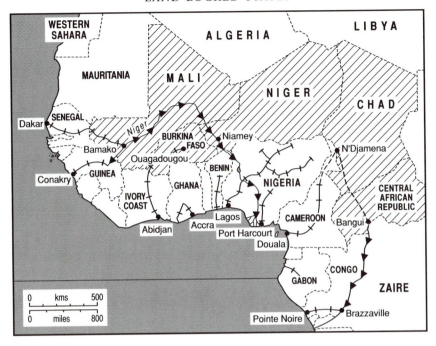

*Map 18* Access routes of land-locked states in West Africa

well. In June 1963 the Mali/Senegal frontier was reopened but the Dakar route never fully regained its dominant colonial position in relation to Mali's external trade.

At the end of 1963 the Federation of Rhodesia and Nyasaland broke up and in 1964 Malawi and Zambia, both land-locked states, achieved independence. Zambia's traditional route to the sea was south through Rhodesia. On 11 November 1965 the settler government in Rhodesia unilaterally declared independence (UDI). Britain reacted to this illegal act by imposing an oil embargo, later extended with United Nations' endorsement to broader trade sanctions. British frigates deployed in the Mozambique Channel prevented oil shipments to Rhodesia via Beira and the oil pipeline to Umtali (Mutare) where Rhodesia's only oil refinery was located. The success was more apparent than real because Rhodesia's oil imports were simply redirected via Lourenço Marques (Maputo) and the Malvernia (Chicualacuala) railway built in 1955. Rhodesia's oil imports continued via this route until March 1976 when the FRELIMO government of newly independent Mozambique closed the frontier to all Rhodesian trade.

Zambia was less fortunate because Rhodesia blocked its oil imports from late 1965. Thus the only real effect of the British blockade of Beira was to prevent Zambia, rather than Rhodesia, getting oil via its traditional route. Emergency sources of supply were necessary for Zambia's vital copper mining industry. Canadians airlifted oil from Nairobi in 44-gallon drums stowed in large transport planes which wobbled over the State House onto the short runway of the old Lusaka airport. A hastily assembled motley fleet of road tankers brought in supplies from Malawi over the (surely misnamed) 'Great East Road'. The wet season morass of the Great North Road to Tanzania was aptly called the 'Hell Run' where tanker lorries plunged into ravines and allegedly were swallowed whole by the bogs that in places passed for roads.

In December 1965 the Rhodesian regime increased its pressure on Zambia, and indirectly Britain, by closing the frontier to coal shipments from Wankie (Hwange) to the Copperbelt. This embargo did not last long as the South African government offered Zambia alternative supplies of coal, though without describing the route it would take. Closure of the Zambesi frontier was contrary to long-term South African interests which sought to retain Zambia within the South African economic and political orbit irrespective of Rhodesian priorities. Nevertheless Zambia was made acutely aware that the southern route was vulnerable and so hastened a reorientation of trade routes which had been planned even prior to UDI.

Zambia sought to modernize the Tanzam route to Dar es Salaam. Between 1968 and 1975 an oil pipeline, tarred road and railway were completed in that order by the Italians, Americans and Chinese respectively. The variety of aid sources itself tells a story. The West was not convinced of the economic viability of a Tanzam railway and in various feasibility studies rejected the project largely because they made wrong assumptions and so asked the wrong questions. Only the Chinese grasped the point that the railway was political rather than economic. Zambia, in attempting to break away from the hegemony of white minority-ruled southern Africa, was making a political decision, not one based projected financial profitability.

The railway, completed on time in 1975, was not the success anticipated and the reorientation of Zambian trade is not complete. The reasons are only partly those predicted by Western aid agencies. The line was constructed cheaply with low carrying capacity due to light rails, tight curves and steep gradients. Although of Cape gauge 3ft 6ins (1.065m), it is operated separately from the rest of Zambian railways, with even a half-mile walk for passengers between stations at Kapiri M'poshi. Most significant has been the low standard of maintenance of permanent way, locomotives and rolling stock. Within eight years of the railway being opened 90 per cent of its locomotives had been cannibalized in an attempt to keep the other 10 per cent in running order. Too little money was available and too little attention was paid to running the railway efficiently. Dar es Salaam port was unable to handle the volume of Zambian trade,

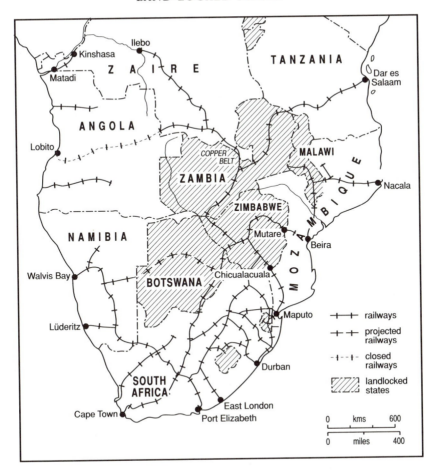

*Map 19* Access routes of land-locked states in south-central Africa

largely bulk copper exports, and no investment capital was available. The net result is that in 1986 about three-quarters of Zambian trade passed through South Africa. East London is the main exit point for Zambian copper, a port half as far again from the Copperbelt as Dar es Salaam, and until 1994 was part of the apartheid state Zambia and Tanzania were both pledged to see destroyed.

Two other rail routes lead from Zambia to the west coast of Africa. The all-Congo (Zaire) route involves trans-shipments from rail to river at Ilebo and at Kinshasa from river to rail to Matadi. The route can not cope with copper exports from Zaire itself, let alone from Zambia, and they are also mainly shipped via East London. The Benguela railway, which branches off the Congo line at Tenke, links with the Angolan port of Lobito. It has not operated since Angola's independence in 1975, closed by the Angolan civil war.

Zambia's continued dependence on the southern route has two main causes: South Africa's stick-and-carrot efforts to maintain the relationship, but more importantly the inability of Zambia and its friendly neighbour states to operate an efficient trunk transport system. Zambia's solution, to press within SADCC for even more alternative outlets to the sea whilst making existing routes viable might be a more effective answer.

Zambia's tormentor, Rhodesia, was itself a land-locked state, and in March 1976 it too was forced to reroute its trade when newly independent Mozambique closed its frontier. This move was anticipated by the Rhodesian regime which in late 1974 completed the long-awaited direct rail link with South Africa via Beitbridge. South Africa had extended its line across the Limpopo as early as 1929 but Rhodesia had neglected to effect a junction, largely to preserve its economic independence from South Africa. Economic survival demanded otherwise. The Rhodesian minority regime did not have long to run, because Mozambique not only closed its border to Rhodesian trade but also opened it to guerillas. The legacy of the direct rail link with South Africa became important, and not altogether advantageous for Zimbabwe after independence.

In April 1980 Zimbabwe became a majority-ruled independent state. But the vulnerability conferred by its land-lockedness quickly became evident when its newly reopened lines of access to the sea came under attack. The Mozambique government was challenged by the South African-backed Mozambique National Resistance (MNR). One tactic of the MNR which hurt the Mozambique government and Zimbabwe, whilst benefiting South Africa, was to blow up the Beira to Mutare railway and oil pipeline and to close the Zimbabwe to Maputo (Chicualacuala) rail route. During these periods of disruption the import and export trade of Zimbabwe had to flow through South Africa via the direct rail link. The Beira and Maputo routes soon acquired a reputation for unreliability, and exporters in Zimbabwe, including parastatal organizations, were easily induced to sign favourable long-term contracts with South African Railways to carry goods out of Zimbabwe via Beitbridge. For much of the 1980s, 85 per cent of Zimbabwe's exports were routed through South Africa, mainly via Beitbridge. South African railways and harbours enjoyed the benefit of the additional traffic and Zimbabwe became politically dependent on South Africa. Zimbabwe responded by deploying troops within Mozambique to keep the Beira/Mutare corridor open. The Chicualacuala railway was closed from August 1984. Again the vulnerability of the land-locked state had been clearly demonstrated and its independence compromised.

In January 1986 the government of Chief Jonathan of land-locked Lesotho was overthrown by a *coup d'état* engineered by South Africa by means of a trade blockade. Lesotho is an enclave, entirely surrounded by South Africa, and so had no alternative means of access. The blockade was enforced because of South African government hostility to the policies of Chief Jonathan, who allegedly gave sanctuary to members of the African National Congress (ANC), and also introduced North Koreans to train special paramilitary units.

Lesotho's new military leader, Major-General Justin Lekhanya underlined the plight of his country in a telexed message to the Secretary-General of the United Nations in which he said:

> The situation is deteriorating rapidly to a point where the safety and security and well-being of Lesotho as a sovereign state are now in jeopardy. We are merely seeking your assistance to address what has become an emergency and a difficult situation unique in the history of our small, land-locked state.

The United Nations response to the blatant bullying of South Africa was limited to 'grave warnings'.

The problems faced by land-locked states are not confined to southern Africa. In 1985 food shipments for Chad were refused passage through Nigeria allegedly because of congestion in the Nigerian transport system, though it was more probably related to a contretemps between the two governments over the border area. Cargoes were diverted to a much longer route via Cameroon. The Chad government was not threatened by this hostile action but it illustrates the vulnerability of land-locked states. In 1979, Uganda was deprived of military supplies and petroleum products imported by rail via Mombasa by the government of Kenya. Partly as a result Amin's government of Uganda was toppled. Few cared about the blockade at the time, indeed many welcomed it, because Amin was overthrown, but again the incident underlines the vulnerability of land-locked states. The border was closed by Kenya, not Uganda, a fact not altered by the praise for the Kenyan action.

Amin, conscious of the weakness of land-locked Uganda, had earlier suggested Uganda be ceded a corridor between Kenya and Tanzania to a port (Tanga) on the Indian Ocean. Before deriding this proposition totally it is as well to remember similar proposals which have been carried out. The *Caprivi Zipfel* was created in 1890 to give access to the Zambesi. If the aim was to navigate that river the Germans had forgotten the Victoria Falls downstream. The *Zipfel* is also impassable in the wet season. Yet it still graces the political map of Africa like a sore thumb. The preamble to the United Nations Resolution of 1950 to federate Eritrea with Ethiopia refers to Ethiopia's legitimate need for adequate access to the sea. That consideration led to a war which lasted thirty years. It arose partly because the United States wanted its then ally, Ethiopia, not to remain the land-locked state it had been until incorporated into Mussolini's East African Empire in 1936.

121

Some land-locked states dream of access to the sea. For example, Hastings Banda made an irredentist-type claim over northern Mozambique, east of Lake Malawi in 1964. Although not followed up, the claim *was* made. The abortive Swaziland/South Africa land deal of 1982 would have given Swaziland access to the sea at Kosi Bay. Swazi aspirations were thwarted by legal action taken on behalf of the KwaZulu 'homeland'. Although the chief attraction of the deal for the Swazi seemed to be the irredentist element, it was when the Ngwavuma/Kosi Bay part of the deal was successfully challenged that the whole deal fell apart, perhaps emphasizing the importance the land-locked state attached to obtaining access to the sea. The 1982 proposal brought the drive for access to the sea of land-locked states in southern Africa full circle. In the 1890s the Transvaal Boers strove to reach the sea, also at Kosi Bay. They were blocked by the British, but a long, narrow finger of the Transvaal pointing towards the coast remains as evidence of thwarted ambition.

Land-locked states are inherently weak and few have alternative routes to the sea. They might plead for special aid, but a case for additional routes to the sea is unlikely to succeed. The best chance of a land-locked state gaining access to the sea is by federation, and then the seaboard state would almost certainly dominate.

# 11

# SECESSIONIST MOVEMENTS

The partition of Africa by the European powers to create colonies which subsequently became independent sovereign states caused all manner of problems for those successor states. Some are land-locked, others are of an odd shape, many have ambiguous boundaries, most have badly located capital cities. Some are too small to be economically or politically viable. Others are too large, particularly in the sense that they encompass disparate groups of people. As states they lack cohesion and unity, and do not easily conform with the nation-state concept that has been imposed on the African polity from the outside. In extreme cases some states have been threatened with break-up by secessionist forces. Civil war has ensued, at enormous cost in lives and human suffering as well as in economic well-being among very poor people in very poor countries.

European spheres of influence and later colonial boundaries were drawn with scant regard to the distribution of African people. Culture group areas were inevitably cut by those boundaries. As the number of culture or ethno-linguistic groups far exceeded the number of colonial units (about 2000:50 or 40:1), and many overlapped with each other, it was certain that most colonies would contain more than one culture group. Zambia, at independence a country of about 5 million people, contained no fewer than seventy-two different culture groups from which create, in the words of Kenneth Kaunda: 'One Zambia, one nation.' It has been estimated that Nigeria, Africa's most populous state, with about 55 million people at independence, contained as many as 395 ethno-linguistic groups. The task of welding these different groups into a single national state, let alone nation-state, would have been daunting even without other contingent factors. For example, *religion*: the Christian religion mainly brought by Europeans (but of ancient origin in Ethiopia) clashing with the Muslim religion diffused by Arabs throughout North Africa from the seventh and eighth centuries and crossing the Sahara from the eleventh century. Or rivalry between Christian Protestants and Catholics, as in Uganda and Rwanda. For example, *localized mineral resources*: making one region rich enough to contemplate autonomy and secession. For example, *outside commercial interests*: eager to give financial support to would-be secessionists in the hope of securing future profits from the mineral wealth. For example, *geographical factors*: the remoteness of potential secessionist regions and the vast distances within the larger African countries. For example, *poor communications and infrastructure*: adding to regional isolation and weakening command structures from the centre. These factors and more were brought to bear on African states often of fragile coherence. Combinations of these factors to varying degrees, always with ethno-linguistic differences present, gave several African states difficult times, some of which are on-going.

The Belgian Congo (Zaire) was the African colony probably least prepared for independence, and Belgium the colonial power probably least prepared for its new status. The immediate result was a chaos of terror and brutality which underlined the fragility of African independence in 1960, exposed the forces of neo-colonialism and emphasized the dangers of disintegration through regional secession.

Also, for the second time in recent history, the Congo caused Africa to become a board on which international rivalries of non-African powers could be played out without danger to themselves.

Zaire is, in area, Africa's third-largest state after Sudan and Algeria, with an area of over 909,000 square miles (2,350,000 square km), about half the size of Europe excluding the former Soviet Union. It is a country of vast distances: Kinshasa (Leopoldville), the capital city, is about 1,000 miles (1,600km) by air from Lubumbashi (Elisabethville) the copper-mining town in Shaba (Katanga) province; Kisangani (Stanleyville) is over 750 miles (1,200km) by air from Kinshasa and 850 miles (1,360km) from Lubumbashi. Distances by surface transport between the three centres are very much greater and the journeys slow and inconvenient because of the use made of river transport and the trans-shipments from river to rail. Kinshasa to Kisangani by river is 1,075 miles (1,720km) and takes over a week upstream; Kinshasa to Lubumbashi is 525 miles (840km) by river plus 985 miles (1,578km) by rail. The road network at independence was primitive, and by 1994 has only progressed to primary. The Congo in 1960 was a place where communications were arduous and painfully slow.

At independence the population of the Congo was under 20 million, divided between many different culture groups, the majority of which spoke Bantu languages. Different groups predominated in each of the three main urban centres and the only cohesion had been provided by the Europeans, mainly Belgians, who numbered about 50,000 in 1960.

The crisis in the Congo was complex in its causes and convoluted in its course. African political development under the Belgians was rudimentary, recent and regional. There was no Congo-wide political party, which laid the country open to regional conflict. Patrice Lumumba from Stanleyville (Kisangani), the most radical and most popular leader, became Prime Minister through pre-independence elections, and Joseph Kasavubu, the

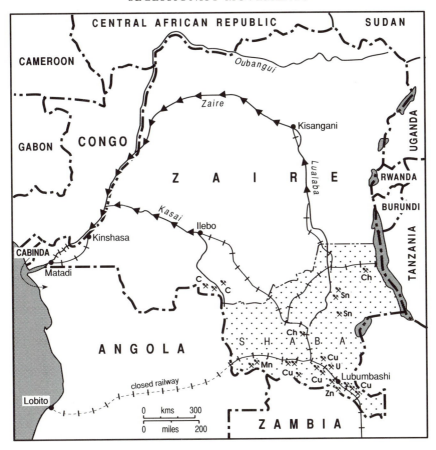

*Map 20* Congo/Katanga

Bakongo leader based in Kinshasa, became president. Independence, brought forward by the Belgians with almost indecent haste, came on 30 June 1960 in an atmosphere of growing distrust between the new government and Belgium. Lumumba fuelled the evolving crisis with a formal but undiplomatic speech on the eve of independence which contained several home truths which the departing Belgians were not pleased to hear.

On 5 July 1960 sections of the Congolese army (*Force Publique*) mutinied against its Belgian officers (there were no Congolese officers). Violence spread across the country and Europeans fled in panic. Lumumba dismissed the Belgian commander of the *Force Publique* and on 8 July appointed a former NCO, Joseph Mobutu, to be Army Chief of Staff and Minister for National Defence. Belgium flew in paratroopers to protect Belgian citizens and property. On 11 July, in mounting confusion, the Katanga (Shaba)

125

region was declared independent by its leader Moise Tsombe. Lumumba appealed to the UN, Ghana and the Soviet Union to help prevent the disintegration of his country. A power struggle developed between the United States and the Soviet Union over direct intervention and eventually a UN-led operation was mounted to enable the Congo to survive as an integral state.

In September 1960 Lumumba and Kasavubu fell out, tried to dismiss each other but were both overthrown by Mobutu in a military *coup d'état*. Lumumba escaped, tried to make his way by road from Leopoldville to his power-base at Stanleyville but was recaptured *en route*, taken back to Leopoldville, then flown to Katanga where he was murdered in January 1961. The Congolese army succeeded in brutally putting down a rebellion in Kasai province but failed to make any impression on Katanga, where Tsombe was able to deploy a white mercenary-officered army. UN resolve to reunite the Congo strengthened but it took two years and the life of Secretary-General Hammarskjöld, killed in a mysterious air crash over Northern Rhodesia (Zambia), to end Katangan secession in early 1963. Civilian rule in the Congo returned under Cyrille Adoula but further rebellion broke out at Stanleyville. Mobutu ousted Adoula and invited Tsombe from exile to the Congo premiership. Katangan forces assisted by Belgians and Americans put down the Stanleyville rebellion and again reunited the Congo. Tsombe won overwhelming popular support but his usefulness was over. In November 1965 Mobutu staged a second *coup d'état*, and Tsombe fled into exile. Mobutu has ruled in Leopoldville ever since.

For almost five years from its independence the Congo was in serious danger of disintegration. Several regions, including Lusambo (Kasai) and Stanleyville staged secessionist attempts but the one that came nearest to succeeding, and would have done so but for a determined effort on the part of the UN, was Katanga.

It had established its own political party, CONAKAT (*Confédération des Associations du Katanga*), mainly among the Lunda people, but then so did other provinces. The political organization was better and the local army was stronger in Katanga than in the other provinces and so were better able to withstand the centrist forces attempting to deny secession. The reason for this was that CONAKAT and the Katangan army were backed financially by the large Belgian mining corporation *Union Minière du Haut Katanga* (UMHK). The Katangan army was officered by white mercenaries and was relatively well–equipped, largely through the financial support of UMHK. The mining company calculated that its interests were best served by having an independent Katanga which it could easily dominate. Tsombe and the Katangan politicians who supported him were conscious of the fact that the Katangan copper mines were the greatest single source of wealth in the Congo and resented having to share their profits with the rest of the country. Copper had been mined in Katanga in pre-colonial times and exported as ingots across Africa. Colonial mining began in 1910 when the spinal railway from the Cape reached the *Etoile du Congo* mine near Elisabethville. Katanga was the northernmost mineral node on the Cape railway, and although it had narrowly escaped the imperialist clutches of Cecil Rhodes, was part of the southern African mining and railway complex. The mineral reserves were considerable and the UMHK had become the largest copper mining company in the world. But the riches of Katanga did not end with copper. Cobalt, uranium and tin deposits combined with copper to make this probably the richest mineral province in Africa after the Transvaal. The stakes were high, and the financial capacity of the mining company to protect its monopoly position was considerable. Katanga was in no way dependent on the rest of the Congo for exporting its mineral products, for two railway routes led from Katanga to the sea without crossing other Congo territory: the Benguela railway which crossed from Katanga to Angola and the port of Lobito, and the Cape spinal railway which crossed into Northern Rhodesia (Zambia) *en route* to the southern African ports.

Katanga was also, as demonstrated above, the region of the Congo geographically most remote from Leopoldville. It lies on the Congo/Zambesi watershed well away from the tropical rain-forest area. Lubumbashi is 4,035 feet (1,230m) above sea level at almost 12° South, compared with Kinshasa at 1,000 feet (305m) above sea level at 4° South. Katanga, whilst contrasting with much of the rest of the Congo, had much in common with the country sharing the broad Congo/Zambesi watershed, Northern Rhodesia. Contiguous and physically similar, in landscape and climate, even the vast copper resources were evenly divided by a fortuitous boundary. Northern Rhodesia was at the time part of the white-settler dominated Federation of Rhodesia and Nyasaland. Mining interests were very important in Northern Rhodesia and the influential mining companies on both sides of the border spoke the same technical, business and political language.

The main culture group in Katanga, the Lunda, was large and dominated the region's politics but had little power nationally. A sense of alienation fuelled the secessionist option, and long survived 1963. In 1977 and 1978 Shaba province was twice invaded, from Angola and Zambia respectively, by 'rebel' forces, some of whom were former members of Tsombe's Katangan army. They were defeated only with outside help, in the first instance by Moroccan and French forces, and the following year, when they made considerable progress by occupying the major mining town of Kolwezi, by French and Belgian intervention.

Secession in the Congo almost succeeded. Factors included the size of the Congo, vast distances, difficulty of communications and the remoteness of Katanga; the political alienation of a distant province dominated by a single culture group; above all the wealth of Katanga, rich enough to go it alone, resentful of having to share its wealth with a distant government it had no major say in, egged on by a well-organized foreign mining company with vested interests. Unity in Zaire remains fragile, with central authority weakened by Mobutu's misrule. Mineral wealth is again a factor: in December 1993 an emotively renamed Katanga declared autonomy, and in mid-1994 diamond-rich Kasai was reported as semi-detached from Zaire.

Nigeria was the next African country to meet a secessionist threat with civil war to retain national integrity. A large country, though territorially only about 40 per cent of the area of Zaire, Nigeria is easily the most populous state in Africa and a country of amazing ethno-linguistic diversity. There are 395 languages in Nigeria, strictly defined as languages and not dialects. Three culture groups predominate: Hausa in the north, Yoruba in the south-west and Igbo in the south-east. The critical step from diversity to divisiveness comes not so much from the enormous number of language groups but, in this case, from the existence of three regionally-dominant culture groups.

Religion is a second major cultural divide in Nigeria, as the north is predominantly Muslim but the south is not. In the north the British practised Lugard's doctrine of the dual mandate and indirect rule, reinforcing the power and Islamic conservatism of the local Emirs. In the south-east and south-west Christian missions made a considerable impact, not least in education. Educational achievements led directly to a westernization and modernization of attitudes which contrasted strongly with the traditionalism of the north.

The third factor in the causes of the Nigerian civil war was oil. The source of Nigerian oil is the Niger delta area, wholly contained within the south-eastern region. The first successful well was drilled in 1956–7, before independence. Throughout the 1960s production was modest, 10–20 million tonnes per annum, but the potential was enormous. After the war ended in 1969 oil production rose rapidly to peak at 111.6 million tonnes in 1974. It stayed at over 100 million tonnes per annum until the world oil glut of the late 1970s when the Organization of Petroleum Exporting Countries (OPEC), the producers' cartel of which Nigeria is a member, set much reduced production quotas. For Nigeria the boom years of the 1970s did not last long enough, but although a cause of the civil war, the oil boom immediately after helped to heal the wounds.

The constitutional history of Nigeria before and after independence reflects concern with the feasibility of creating a unified state by balancing the regional interests of north, west and east, the power bases of Hausa, Yoruba and Ibo respectively. In the run-up to independence in 1960 the emphasis was on finding a constitutional formula that would effect unity through regional balance. During tortuous negotiations all three regions threatened secession but strangely, in the light of subsequent events, the north was thought to pose the greatest threat to Nigerian unity. It was the largest and most distinctive of the regions and fiercely defended its

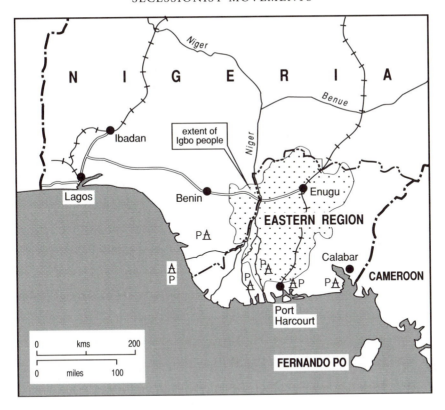

*Map 21* Nigeria/Biafra

traditions and culture. The independence constitution of 1960 was federal, with the three regions (plus Lagos) retaining a high degree of autonomy but delegating certain powers, including control of the army, police, customs and excise, currency, central banking and international trade, to the centre. The principal political parties were increasingly regionally based. Nevertheless, independence was achieved on the basis of a united federal Nigeria. The first federal Prime Minister was a northerner and Muslim, Alhaji Sir Abubakar Tafawa Balewa, deputy leader of the Northern People's Congress (NPC) which controlled the Northern Region government. Balewa ruled Nigeria with support of the eastern-based National Council of Nigeria and the Cameroons (NCNC) under Nnamdi Azikiwe which controlled the Eastern Region government. Excluded from the Federal Government were the Yorubas of the Western Region, whose Action Group (AG) became the official opposition. The north/east coalition did not last long. An AG split in the west led the federal government to take over the regional administration and later to install in the west a minority faction aligned with the north. The controversial 1963 census and the 1964 elections strained north/east relations. Against this unstable and increasingly violent political background, in January 1966 came Nigeria's first military *coup d'état*, which killed Balewa and other leaders of the north and west. The new military leader, General Aguiyi Ironsi, proclaimed a unitary state, which caused further unrest. In July, after anti-Ibo riots in the north, Ironsi was assassinated in an anti-Ibo military *coup* that installed Colonel Yakubu Gowon as new head of state. Gowon was a Muslim but not a northerner, coming from Plateau State in the Nigerian middle belt between the three ethnic power-blocks. Reconciliation with the Ibos, who refused to accept Gowon's authority, failed as thousands more Ibo residents in the north were massacred in September and October 1966. They had been part of the Ibo diaspora, filling jobs in government and administration that required Western-type education not widely provided under colonialism in the Muslim north. Gowon proposed to divide Nigeria's four regions into twelve states, effectively to place the oil resources in the hands of non-Ibo minorities within the existing Eastern Region.

In May 1967 civil war began as the Ibo east under its military governor Colonel Ojukwu declared Biafra an independent state. After an initial Ibo advance through the mid-west region had been repulsed, the military outcome was never in doubt. The tragedy was that the federal forces were unable to end the affair quickly, largely because outside powers, including France, Portugal and South Africa, sustained Biafra until it dwindled to a single airstrip. Sympathy for Biafra was not limited to neo-colonial forces and Gabon, Ivory Coast, Tanzania and Zambia all 'recognized' the secessionist state. Two million civilians died as Biafra was starved into submission and many more suffered severely in two-and-a-half years of war. World-wide television coverage recorded the agony of Biafra in graphic detail.

Superior forces and a determination on the part of the federal government to preserve the integrity of Nigeria made the end inevitable. Biafra gave up the struggle in January 1970. The war over, Gowon pursued a policy of national reconciliation, aided by the boom in oil prices which gave Nigeria a period of unique prosperity. Despite the cost in lives and resources there is little doubt that the interests of all Nigerians and all Africa were best served by the preservation of a united Nigeria.

One underlying aspect of political instability in Nigeria, outwardly expressed in military *coups* and corruption, is the inability to get away from the fault lines between north, west and east. The administrative divisions which were part of the colonial inheritance from the British are still to be found on Nigeria's political map. The basic division of north, west and east created by the British in 1939 survived independence. Sub-divisions of the original three regions occurred in 1963, 1967, 1976 and 1991. But the old regions are still identifiable and if there is to be any hope of ending rivalries based on that split, a radical geographical restructuring of administrative units is needed with new lines cutting across the old regional boundaries. The 1993 constitutional crisis in Nigeria had its critical elements of *déjà vu*. The elections of June 1993 were won to most observers' satisfaction by a Muslim westerner, Moshood Abiola. General Ibrahim Babangida, the northerner military dictator, having carefully vetted the candidates (both multi-millionaire businessmen), and specially created two synthetic political parties for the civilian presidential and national assembly elections, could not bring himself to accept the result. It was widely suggested that this was due to Abiola being a Yoruba. In August 1993 Babangida installed a new unelected civilian government headed by Ernest Shonekan which was more acceptable to the military, but it was soon overthrown in yet another army *coup d'état* led by former defence minister General Sani Abacha. It is too early to say to what extent the old divisions were the main cause, rather than, for example, the fear of exposure of past corruption, but they were present. The fault lines were present before colonization, with the important difference that they then separated political entities. The British, in throwing a boundary around the territory which became Nigeria, ignored the inherent disparities and, through differential administration within the colony, gave the fault lines added emphasis. They survived independence as a particularly destructive part of the colonial inheritance and have persistently prevented Africa's most populous and potentially richest colonial creation, Nigeria, from attaining prosperity and political stability.

The Sudan attained independence from the Anglo-Egyptian Condominion in 1956. It is, in area, the largest country in Africa with 1.35 million square miles (2.5 million square km). With this vast size comes ethnic diversity. One authority quotes nineteen major ethno-linguistic groups and 597 distinct sub-groups. Superimposed on this complex ethnic pattern are two broad areas of strongly contrasting cultures, the Muslim north and the Christian and animist south. The Muslim north was the larger and stronger and the south lived in fear of Muslim domination even before independence.

Egypt, formally co-ruler of the Sudan with Britain, wanted to absorb the northern (Muslim) Sudan into the Egyptian state. The southern Sudan favoured separation from the north or at most federal association with it. Britain believed that parliamentary democracy was the answer to all the potential constitutional problems and that view prevailed. For the first time in post-war Africa, but certainly not the last, the sophisticated, checked-and-balanced, British-made constitution at independence proved inadequate for the task of providing a durable structure of political stability. In 1958 the government and constitution was overthrown in a military *coup* led by General Ibrahim Abboud.

Before this, even before the British withdrew in 1956, a civil war had started between the minority south, fearful of Muslim domination, and the majority north, determined that the will of the majority should prevail throughout the country. It lasted for seventeen years, cost as many as 1 million lives and outlasted the military dictatorship of Abboud and the successor civilian governments of 1964–9. A second military *coup d'état* brought General Nimeiri to power in 1969, who in March 1972 achieved agreement with the southern rebels to end the civil war on the basis of regional autonomy for the south. The fragile peace was not to last.

One reason for the north ending the war and granting regional autonomy was that oil had been discovered in the south, which would greatly help the national economy. The south's oil was exploited and little given in return. There was little economic development in the south and shortages in basic consumables sparked riots throughout much of the country. Nimeiri accepted massive aid from Saudi Arabia, who in return asked that Muslim Sharia Law be applied in the Sudan. The south's fear of Muslim domination was accentuated; their oil resources was developed mainly for the benefit of the north and their other great resource, water, was also in danger of being seized, also mainly for the benefit of the north in the Jonglei barrage and canal scheme, which would have diverted the White Nile away from the Sudd to increase the supply of water available for irrigation downstream in the north. The final straw was a proposal from Khartoum to divide the south into three provinces. The southern leaders interpreted it as 'divide and rule'.

In 1983 Colonel John Garang, a southern Christian, set up the Sudan People's Liberation Army (SPLA) and secessionist civil war recommenced. The national government in the north changed several times but always remained in control whilst moving steadily towards Muslim fundamentalism, including the Sharia Law, complete with floggings, amputations and executions. The first six years of renewed civil war cost 500,000 lives. Many more died of the famine which ravaged the south as the war thwarted emergency relief attempts. The northern army devastated wide areas of the south deliberately as a weapon of war, so intensifying the famine. Providing food relief for the war-torn southern Sudan is extremely hazardous; news reports are at best sporadic and have to compete with more dramatic crises for international impact.

There is no permanent end in sight to the civil war in the Sudan. Secession of the southern Sudan is a possible outcome, the likelihood of which changes from one season of fighting to another. The two parts of the country are very different and the aggressive imposition of a harsh version of an alien culture by the majority on a culturally distinct minority, combined with exploitation of the resources of the minority area by the majority, holds little prospect for peace. The Muslim majority in the Sudan, in control of the army, internal politics and the economy, see it as an internal affair of state and brook no outside interference on the side of the south. Meanwhile the north draws heavily on Saudi Arabia for support,

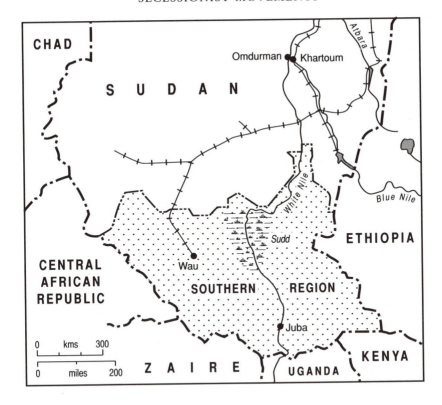

*Map 22* Southern Sudan

which has cultural strings attached, but strings which are perfectly accept-able to the faction controlling the Muslim majority. The geographical remoteness of the southern Sudan exacerbates an already desperate situation. It is a completely land-locked area, nowhere less than five hundred miles from the sea with virtually no access routes alternative to those from the north, as the various aid agencies have found to their cost. What is more, the southern Sudan is surrounded by states which themselves have recently experienced political violence, unrest, and civil war, notably, Ethiopia, Uganda and Chad.

Similarities between the experiences of the Congo, Nigeria and Sudan are interesting. All are large countries, in area if Sudan (2,505,813 square km) is 100, Congo 94 and Nigeria 37. Each would-be secessionist region was remote and mineral-rich; the least remote, Biafra, being the richest in mineral resources. Each state has wide ethnic variation with the ethno-linguistic group dominating the secessionist region different from that controlling the centre. In all three cases at the centre and the region there

133

were military regimes. In the case of Katanga and Biafra there was overt outside support for the secessionist region from countries which sought to benefit economically or politically. A religious divide exists in Nigeria and Sudan but not in Zaire. Religion is much more crucial an element in the Sudan than in Nigeria, with external influences evident. The sophisticated political structure and experience developed and accumulated over a long period of time in Nigeria contrasted with the lack of such structures and experience in the Congo. In the Congo, Nigeria and Sudan the threat of secession was present before independence, and in all three attempted secession was a crisis waiting to happen. The divisions came along well-known fault lines which had been evident before independence.

Having created the problems by putting boundary lines around disparate groups, colonial rule often exacerbated the problem by treating the groups differently. At independence the colonial powers failed to cope with antici-pated problems. The British in the Sudan, Nigeria and elsewhere, drew up independence constitutions of elaborate checks and balances, backed by the Westminster parliamentary model. The swift destruction of successive tailor-made independence constitutions did not shake British faith about the appropriateness of the Westminster model for Africa. The British sin of misplaced but well-intentioned commission has to be put alongside the Belgian sin of blatant omission and the French off-the-peg treatment of one and all. In no way did the colonial powers smooth the passage beyond independence, another part of Africa's colonial inheritance.

There are other states in Africa which have faced, or face, the threat of secession from a region, for example, Zambia, Senegal and South Africa. The Barotse province of western Zambia was indirectly ruled by the British as a separate protectorate. It is culturally distinct, with its own political history which predates colonialism. Zambia is a state of great ethno-linguistic variety. Barotseland is geographically remote within Zambia, though not on the scale of the Sudan or Zaire, Mongu being about 380 miles (610km) from Lusaka. Secessionism has never progressed beyond Lozi rumblings of discontent with the centre. Perhaps above all Zambia's unity is best preserved by the geographical centrality of its Copperbelt and the fact that the two 'wings' of the country have no great resources.

The Casamance region of Senegal is remote from Dakar: Ziguinchor is 360 miles (580km) away. More importantly the Gambia, and the Gambia River, which is crossed by ferry, intervenes. Until the Trans–Gambia highway was built, the road distance between state capital and regional capital via Tambacounda was about 515 miles (825km). In the early 1980s the people of the Casamance province, mainly of the Diola culture group and distinct from the Wolofs and Mandinkas of both the Gambia and northern Senegal, protested at what they regarded as the neglect of their region by Dakar. The leaders of the Casamance were imprisoned and the southern region's two provinces were sub-divided into four, a tactic reminiscent of Nigeria and Sudan, to weaken organized opposition in the periphery. The action taken in Dakar decisively headed off the secessionist movement, though discontent still rumbles on. A major oil strike in the Casamance could transform the situation.

South Africa was created in 1910 as a white minority-ruled union of four former British colonies, two of which had been Boer republics. Three of the former colonies were given a formal share in the Union government through the device of geographically separating the capital city functions between administrative (Pretoria), legislative (Cape Town) and judicial (Bloemfontein). Only Natal missed out in this sharing of functions. Always miffed at taking second place to the Cape in nineteenth-century British administration, more British than the British, especially when comparing themselves with the Boers, and the most reluctant to enter the Union, many whites in Natal have long harboured romantic thoughts of going it alone. As South Africa moved towards majority rule Mangosuthu Buthelezi of KwaZulu and the Inkhata movement also developed secessionist tendencies. The first openly-made overtures to the whites of Natal came in the *Indaba* (conference) of 1986. The response was cautiously favourable. Buthelezi later moved to ally himself with extreme right-wing whites who nurtured secessionist ideas of a 'Boerstan' arising from a partitioned South Africa. In the event Inkhata belatedly participated in the April 1994 election and controversially gained a majority in KwaZulu/Natal. Buthelezi accepted the Home Affairs portfolio in the Mandela interim government but no great reconciliation has yet taken place. The possibility of Buthelezi being tempted to play the secessionist card and attempt to take KwaZulu/Natal out of South Africa remains. The longer the delay the less the threat, but the ability of the ANC to hold South Africa together will be its greatest test. Any perceived weakness in that area is likely to be exploited to the full.

Putting lines around areas on the map of Africa in an almost arbitrary manner, making the areas so defined colonies and then sovereign states, was a process almost certain to cause problems. Ethnic diversity; other cultural differences, including religion; centrist politics, designed to overcome centrifugal forces but often overdone; great areas; vast distances; unequally distributed resources; perceived unfairness in regional budget allocations are only some of the problems that African states have to grapple with. Certain combinations of these factors can lead to secessionist movements and things begin to fall apart. To some extent it is inevitable that there would have been such problems, no matter how Africa were divided into fifty or even fewer sovereign states, but the particular colonial experience which made the inevitable actual is responsible for much of the present turmoil in Africa.

# 12

# IRREDENTISM

The word irredentist was introduced into the English language from Italian politics in 1882 to describe: 'an adherent of the party which advocates the recovery and union to Italy of all Italian-speaking districts now subject to other countries' (Shorter Oxford Dictionary). By extension irredentist is the term applied to any group trying to unite under one flag all districts where any particular language is spoken; their policy is irredentism. Because irredentism presupposes a sovereign state of some linguistic homogeneity, there are few examples in Africa; the African state is generally characterized by ethno-linguistic variety. Somalia, Lesotho and Swaziland are African states which are largely ethno-linguistically homogeneous. They have shown very different attitudes towards irredentism but it is a potentially dynamic and emotive force which in certain circumstances can shatter the political *status quo*.

Somalia is one of Africa's poorest countries, yet it has chosen for much of its existence to sacrifice economic progress for the ideal of Somali self-determination and to try to extend the Somali Republic to unite all Somali speakers in the true spirit of irredentism. Its irredentist policies have set Somalia at odds with its neighbours and have seen more money spent on guns than on butter. When that unifying ideal has not provided the driving force in national politics, the state of Somalia has degenerated into internecine warfare between the Somali clans. In the 1990s civil war has reached such a pitch as to threaten disintegration of the state itself and has caused untold suffering among ordinary people.

At independence in 1960 Somalia was already a rarity in Africa, a union of two former colonies, British Somaliland (68,000 square miles/174,000 square km, 650,000 people) and Italian Somaliland (178,000 square miles/456,000 square km, 1,230,000 people). Even then almost 1 million Somalis lived outside the newly-formed Somali Republic, occupying areas of Ethiopia, Kenya and Djibouti (French Somaliland), totalling about 128,000 square miles (328,000 square km). Over one-third (34.7 per cent) of all Somali-speaking people and over one-third (34.2 per cent) of all land occupied by Somali-speaking people lay outside the Somali Republic. Whilst

136

other African states have struggled to create a sense of national unity among people of many diverse ethno-linguistic and culture groups, Somalia has been absorbed for much of the time with the aim of uniting all Somali-speaking people under the single flag of a 'Greater Somalia'.

Somali irredentism, and its total rejection by the neighbouring states containing Somali minorities, has been an intractable problem since Somalia's independence. It has been seriously exacerbated by the perceived strategic position of the Horn of Africa, which has attracted the attention of imperial powers for a century and a half. Thousands have been killed in the cross-border warfare that has punctuated the post-colonial period, thousands more have died of related causes and literally millions have become refugees. When, between bouts of irredentism, Somalia has turned in on itself, warfare between the Somali clans has been even more bitter and every bit as destructive, as warlords with well-equipped private armies have fought each other for political supremacy. Recently it has seemed that almost all Somali leaders have had their fingers pressed hard on a self-destruct button, as all outside efforts to help ameliorate a desperate situation have been rudely brushed aside. The hand of the UN, albeit mainly Pakistani and American troops untutored in local clan politics, extended with humanitarian emergency food aid, was well and truly bitten in 1993–5. All this in an area of awful natural poverty, a drought-stricken tract where, under the most favourable of local conditions, it is difficult to eke out a living.

Over two-thirds of Somalis are pastoral nomads living in a harsh semi-arid environment. The migratory patterns of such a people present a dynamic force inevitably at odds with fixed international boundaries, especially so when those frontiers are for the most part arbitrarily drawn straight lines. To make the irredentist problem more complex, for over a century the nomadic Somalis are known to have migrated westward: in the south across north-eastern Kenya, displacing other groups as they moved; in the north into Djibouti. The Ogaden area of eastern Ethiopia is criss-crossed by well-defined Somali seasonal migrations. This westward drift of Somali-speaking people has steadily extended the districts they occupy and has thereby increased the area claimed by Somali irredentism.

During the European partition of Africa, the Somali coast was divided between the spheres of influence of Britain, France and Italy. The British and French were first interested in the area in the nineteenth century when the Suez route to the East was developed, first with an overland section from Alexandria to Suez, and after 1869 via the Suez Canal. Aden and Obock (and later Djibouti) were developed as coaling and naval stations by the British and French respectively. To supply food for Aden, with its barren hinterland, the British extended their influence over the north Somali coast across the Gulf of Aden. Italy, for so long occupied with its own unification and irredentism, was a late player in the great game of scramble.

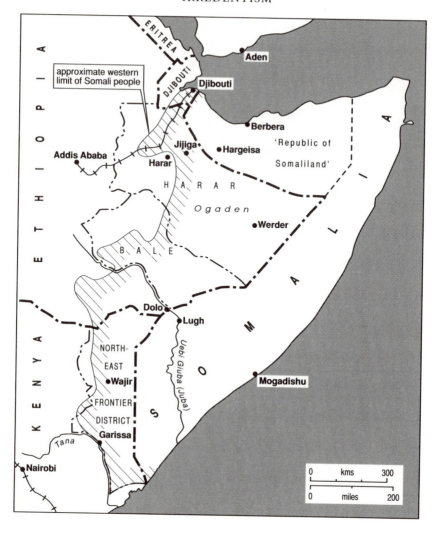

*Map 23* Somalia

The parts of Africa not already claimed by one of the other European powers were the least attractive. They included the Somali coast between British East Africa and British Somaliland, and the Red Sea coast between French Somaliland and the Sudan which in 1890 became the Italian colony of Eritrea. Italian hopes of a great East African Empire linking Eritrea and Italian Somaliland foundered when they failed to conquer Abyssinia (Ethiopia), being beaten at the Battle of Adowa in northern Ethiopia in 1896. In 1924 Britain ceded to Italy a large piece of British East Africa west of the Juba river, Jubaland, an area of 36,740 square miles (94,050 square km) entirely inhabited by Somalis. It was Italy's 'reward' for entering the First World War against Germany. In the secret Treaty of London of 1915 the British and French had agreed in principle that should they increase their colonial territory in Africa at the expense of Germany, Italy might claim 'some equitable compensation'. Apart from fulfilling an agreement between the two imperial powers the Treaty of 1924 went some way towards solving the Somali minority problem in Kenya by substantially decreasing the Somali districts in the colony, though that was not its main aim. In 1935 the Italians, now under the fascist dictatorship of Mussolini, set about realizing the dream of an East African Empire that had been shattered almost forty years before. Ethiopia was conquered and the Italian East African Empire created, comprising Ethiopia, Eritrea and Italian Somaliland. It lasted a mere five years.

During the Second World War British and Commonwealth forces defeated the Italians and occupied all of the former Italian East African Empire. Ethiopia was claimed by its Emperor, Haile Selassie, who quickly returned from exile in Europe. Britain returned Ethiopia to independence with the exception of the Ogaden region, the eastern part of Ethiopia mainly inhabited by Somali-speaking people. Britain also administered Eritrea and Italian Somaliland both during and after the Second World War, until the newly-formed UN could decide what should be done with the territories. In 1946 the British foreign secretary, Ernest Bevin, proposed the creation of a 'Greater Somalia' embracing the vast majority of all Somali-speaking people. The plan proposed uniting British Somaliland, Italian Somaliland and the Ogaden region of Ethiopia, all of which were then under British Military Administration, in a single colonial territory. If accepted the plan would have taken the sting out of subsequent Somali irredentism, although it would have left Somali minorities in Kenya and French Somaliland. But accepted it was not. The United States and the Soviet Union, suspicious of British intentions in the strategic Horn of Africa, rejected the proposals out of hand. After protracted deliberations in the UN in 1950 the Italians were handed back their colony of Italian Somaliland as a UN Trust Territory. In 1955, with reluctance, the British allowed Ethiopia to reoccupy the Ogaden. A great opportunity to foresee and prevent the Somali irredentist problem had been lost through foreign power rivalry.

Between the independence of Somalia in 1960 and that of Kenya in 1963 the British were pressured to cede the remaining Somali areas of Kenya to Somalia. A new North-Eastern Frontier District was created by the British in Kenya in 1963 immediately prior to an Anglo-Somali Conference in Rome. The extent of the new district almost exactly coincided with the westward spread of the Somali-speaking people. It is possible that the British intention in creating the new district was to pre-define the limits to which they were prepared to go in meeting Somali demands, and perhaps to indicate that the new district was dispensible. The Somali negotiators stood out for the whole of the former Northern Province of Kenya, which was not acceptable to Britain. No agreement was reached and the talks ended in deadlock. Another opportunity was lost to ease the Somali irredentist problem, this time probably because the Somalis were too greedy in wanting territory far beyond that occupied by Somali speakers.

In December 1963 Kenya achieved independence with the 1924 Jubaland boundary with Somalia intact, and with a substanial Somali-speaking minority, mainly in the North-Eastern Frontier District, but also in Nairobi. In July 1964 Somalia, alone with Morocco, refused to sign the Cairo accord of the OAU by which member states agreed to respect the colonially-inherited boundaries of Africa. The Somali government could not agree to the boundary *status quo*, which would have been tantamount to renouncing their irredentist policy.

Kenya's independence constitution, negotiated with the British, specifically protected regional and minority, including Somali, interests. However, as so often was the case with these painstakingly constructed and carefully balanced constitutions, it was swept aside in the constitutional reform of 1965. A guerilla war developed along the Kenya–Somali border which cost many lives and great suffering but did nothing to advance the cause of Somali self-determination.

The Somali irredentist cause received a further set-back in French Somaliland in 1967 when a referendum declared against independence, which had been seen as a first step towards integration with Somalia. The carefully massaged vote found almost 60 per cent in favour of continuing ties with France rather than independence. Many Somalis had been expelled and others imprisoned before the referendum and the anti-Somali vote was assiduously managed. The French had shown their hand in 1966 when they changed the name of the overseas territory from *Côte Française des Somalis* to *Territoire Française des Afars et des Issas*, a more cumbersome title, but one that accurately indicated French political intentions by spelling out the names of the tribes and avoiding use of the word 'Somali'. Again Somali aspirations were thwarted by the intervention of an outside power, this time a France anxious to maintain its military and naval base at the strategically important southern entrance to the Red Sea.

The super-powers, having blocked British plans for Somalia, were not long in establishing themselves in the strategic Horn of Africa. In return for a Red Sea base the United States helped Haile Selassie acquire Eritrea where the base was located, and provided him with vast quantities of arms to help keep his fragile Ethiopian empire together. The Soviet Union outbid the West to supply Somalia with weapons in return for a base at Berbera on the the north Somali coast of the Gulf of Aden. Between them the super-powers flooded Ethiopia and Somalia with modern weaponry which the two countries eagerly used in the sporadic warfare between them during the 1960s.

Late in 1967 Prime Minister Egal's new Somali government recognized the impoverishing effect and apparent futility of pursuing irredentism and attempted to negotiate a settlement with Kenya and Ethiopia. The guerilla war died out and Somalia was, almost for the first time, at peace with its neighbours. But removing the unifying external issue led to internal political strife which culminated in the military *coup d'état* which brought Siad Barre to power. Egal's policies were continued, with emphasis placed on trying to solve Somalia's formidable internal development problems. However, the Ethiopian *coup d'état* of 1974 which overthrew Haile Selassie and the hype of publicity which accompanied Siad Barre's chairmanship of the OAU once more brought Somalia's irredentist ambitions to the fore. The Soviet Union, Somalia's super-power ally, now replaced the United States in the counsels of revolutionary Ethiopia and attempted, without success, to mediate between the two sides on the issue of self-determination for all Somali speakers.

Somalia was in no mood for negotiation and moved to take advantage of the chaos in Ethiopia. In the Ogaden the Western Somali Liberation Front was formed to fight for the absorption of the Ogaden into Somalia. By the latter half of 1977 full-scale war was in progress with Somali army joining in. The Somalis were initially highly successful and pushed the Ethiopian army out of the Ogaden. It might be argued that the Somalis pushed too far west, well into Galla country and threatened the Ethiopian core. The Ethiopians regrouped and, equipped by the Soviet Union and 'advised' by 15,000 Cubans, inflicted a crushing defeat on the Somalis at Jijiga in March 1978. At a single blow the formal war was virtually over, but Somali guerilla activity, Ethiopian air-raids and Somali cross-border incursions rumbled on into the early 1980s. The Somalis broke with the Soviet Union because of the part they had played in support of the Ethiopians and the super-power carousel took another turn. Certainly the Somalis again suffered from the super-powers taking such a close interest in the affairs of the Horn of Africa.

The war coincided with serious drought, and the two factors combined to drive the Ogaden Somalis from their traditional pastures. Internationally-funded emergency refugee camps were set up along the Somali–Ethiopian frontier as feeding and watering points. Emergency food supply lines were insecure and health hazards in the crowded unhygenic camps multiplied with the drought. All around flare-ups of the war caused upsurges in the number of refugees and further impeded relief work. The Somali government was threatened by the severe drain on its resources. The guerilla war eased and by 1982 Ethiopia had reasserted its full control over the Ogaden. Ethiopian forces then mounted cross-border raids on Somalia which now turned to the United States for military supplies and political support. The super-powers so completed their musical chairs in the Horn, each moving to support the side opposite from the one they started with.

A stalemate developed between Ethiopia and Somalia as the former became preoccupied with the more threatening Eritrean and Tigray problems, whilst the latter turned to pressing internal concerns. Siad Barre took more dictatorial powers and ran into increasing opposition. He survived an army-led attempted *coup d'état* in 1987 only to face growing inter-clan rivalry among the Somalis which eroded his power. After bitter fighting towards the end of 1990 the government of Siad Barre was finally overthrown in early 1991. An interim government was set up under Ali Mahdi Mohamed but the fighting did not cease and Somalia seemed submerged in a surge of internecine warfare. The former British Somaliland declared itself independent from the rest of the country as the Republic of Somaliland but was not recognized by the rest of the world, indeed was hardly noticed. In and around Mogadishu, the capital, inter-clan fighting was particularly fierce through much of 1991 and 1992. Foreign nationals were evacuated and the badly needed international aid organizations were forced to flee. Food aid and other supplies were unable to get through even by sea as the port of Mogadishu was continually under fire. Beyond the capital there was also fighting and even in the self-declared Republic of Somaliland inter-clan warfare bubbled to the surface. As a background to all this, ordinary people, ravaged by drought and war, were starving to death. Belatedly the UN decided to get involved, in effect to force-feed Somalia. To get the food into the country and then to distribute it to the people in need required force. In late 1992 the United States and Pakistan provided the first troops for a UN operation which proceeded throughout 1993 and 1994 in the teeth of opposition from Somali clans still vying for power, now based on control of emergency food supplies.

An end to the fighting, although necessary, would only be the first step in attempting to solve the problems of Somalia. Reports spoke of enormous damage to almost all structures in and around Mogadishu. People died in their thousands because food and medical supplies could not get through. Basic services, including urban water supplies, were destroyed. Beyond the crippled city drought gripped the countryside and thousands were at risk of starving to death with little hope of food aid. The basic infrastructure of the country, such as it was, has been severely damaged. As part of ending the self-destructive inter-clan rivalry the rift with the secessionist north remains to be healed. In March 1995 the UN withdrew from Somalia with little to show for an expenditure of $3 bn and hundreds of lives.

Beyond Somalia itself the basic problem of Somali irredentism remains unsolved. Super-power suspicion prevented an early solution in 1946. Super-power rivalry exploited Somali aspirations and Ethiopian disarray. Super-power arms brokers supplied the weapons for the Somalis to shoot off their own feet. In the post-cold war situation the Horn of Africa has a much diminished world strategic significance. Ethiopia in the 1990s is not the imperialist state it recently was, as the evidence of an independent Eritrea shows. Perhaps when, with outside help rather than direction, Somalia sorts itself out internally, the irredentist problem might be approached afresh in a more conciliatory, less confrontational manner. Somali irredentism is a problem that will not go away and so is better faced squarely in a realistic, common-sense way not only by Somalia but by Ethiopia, Kenya and Djibouti (and France) as well.

The southern African kingdoms of Lesotho and Swaziland were the products of the early nineteenth-century *Mfecane*, a great upheaval among the African people of the area with its epicentre in Zululand but its effects felt as far afield as present-day Zambia, Malawi and Tanzania. Probably initially sparked by over-population in Zululand, fighting between various clans in the area led to much bloodshed, famine and the large-scale displacement of many groups of people throughout much of the sub-continent. On the positive side the *Mfecane* saw the creation of several new 'nations', the Zulu under Shaka at the epicentre, the Ndebele under Mzilikase who eventually settled around Bulawayo in present-day Zimbabwe after the intervention of the Boers, the Basuto under Moshoeshoe in the mountains of present-day Lesotho, and the Swazi under Sobhuza who settled across the Pongola river from the Zulu in an area based in present-day Swaziland.

The *Mfecane* immediately preceded the occupation of the South African high veld and Natal by white settlers, mainly the trek-Boers emigrant from British liberal rule in the Cape Colony. The Boers clashed with the new African kingdoms. The Ndebele who were settled in the present-day western Transvaal when the trek-Boers arrived fled in 1838 after a series of skirmishes and battles to present-day Zimbabwe, where they established a hegemony over other African groups that lasted until they were overcome by white settlers in Rhodesia in 1893 and 1896. The Basuto and Swazi were also harassed by the Boers, who sought ever more land for their herds and protection from cattle-raiding and rustling by rolling back the frontiers of their republics, forcing the African kingdoms into the less desirable margins.

Increasingly the Basuto, who had only recently been forged into a 'nation' by Moshoeshoe from various groups desperately fleeing the Zulus, were forced back into their mountain fastnesses by the Boers. They quickly learned about the political and ethno-linguistic differences that existed between their persecutors and the British. Moshoeshoe cleverly exploited those differences and in 1868, after a series of set-backs against the Boers, appealed to the British for protection. The British responded and a political boundary between the Basuto and the Boer Republic of the Orange Free State was established. The stability this brought had its disadvantages as many areas previously under Basuto control passed to the Boers. Notably these areas included the upper Caledon valley, by far the most fertile part of the Basuto

Kingdom. This 'Conquered Territory' is still a matter of dispute, the Boers asserting it is theirs by right of conquest, the Basuto claiming it is theirs by historical right. With the land taken by the Orange Free State went Africans, many of whom were absorbed into Boer farms. Today there are many South Sotho living in the Orange Free State, particularly in the Bantustan 'homeland' of QwaQwa, of the same ethno-linguistic group as the inhabitants of Lesotho, and the potential cause of an irredentist claim from Lesotho.

From 1868 Lesotho (Basutoland) was ruled by Britain as a protectorate, along with Bechuanaland (Botswana) and Swaziland. This was a variation on the British practice of indirect rule in which the hand of Lugard, who served in Bechuanaland, can again be seen. The High Commission Territories, as they became known, were never allowed to be absorbed by South Africa, despite strenuous efforts by the Union government in the inter-war period. Lesotho and Botswana achieved independence in 1966 and Swaziland in 1968.

Although Lesotho has claimed the 'Conquered Territories' it is not primarily an irredentist claim. That is not surprising because the greatest concentration of South Sotho in South Africa is in the artificially created apartheid 'homeland' of QwaQwa where the population has been squeezed into a small area bereft of resources. QwaQwa was the least convincing of South Africa's 'homelands'. Some 200,000 people occupy about 239 square miles (620 square km) of uninviting high veld into which the majority were forceably 'removed' or 'resettled' under apartheid. Less than 10 per cent of the land is arable, whilst 80 per cent is officially termed 'grazing'. Fewer than 10,000 jobs are available within the homeland, the overwhelming majority being government jobs in the tertiary sector. In the mid-1980s its Gross Domestic Product (GDP) was R110 million, just 30 per cent of its Gross National Product (GNP) of R363 million. In other words, more than two-thirds of its income was derived from outside its boundaries, almost all from white South Africa. There is simply nothing in QwaQwa to whet any irredentist's appetite.

Swaziland was also formed in opposition to the Zulu in the early part of the nineteenth century and went on narrowly to escape annexation by the Boers in the 1890s. About half the area of Swaziland was alienated to mainly Boer settlers during the colonial period, and the boundaries drawn by the Boers and the British in the 1880s placed in the Transvaal a great deal of territory traditionally occupied by the Swazi. That territory contained large numbers of Swazi who, down to the present time, owe tribal allegiance to the King of Swaziland and form the object of any irredentist ambition on the part of Swaziland.

Under the delusion of Grand Apartheid the South African government also set up a Swazi 'homeland' in South Africa, KaNgwane. Larger than QwaQwa, it comprised two separate blocks of territory, Nkomazi, contiguous with Swaziland, and Nsikazi, a detached enclave completely surrounded by white South Africa north of the Delagoa Bay railway. Together the two pieces of KaNgwane totalled 1,486 square miles (3,850 square km) in 1982. The total population of the KaNgwane homeland was about 854,000 in 1982. This included two basic elements, those actually living in the homeland, the *de facto* population, and those who were under the laws of apartheid were held to live there, but who actually lived elsewhere in South Africa, and were added to the *de jure* population. Both elements contained some non-Swazi people. At the same time the population of Swaziland itself was about 650,000, so that there were more ethnic Swazis resident outside Swaziland than inside.

In 1982 the South African government proposed a land deal to Swaziland. In return for signing a secret Defence Accord, among other things banning ANC activity, Swaziland was to be given the homeland of KaNgwane and all its people, plus another homeland area, Ngwavuma, consisting of 1,595 square miles (4,132 square km) and 96,000 people, which importantly would give land-locked Swaziland access to the sea at Kosi Bay. Had the deal gone through it would have been the achievement of an irredentist ideal *in extremis*, with the smaller swallowing the larger.

The main South African aim in proposing the deal was to ensure security from possible ANC infiltration via Swaziland by a defence pact. A similar, more publicized defence agreement, the Nkomati Accord, was signed with Mozambique in 1984. The secondary aim was to 'give away' almost a million blacks, so slightly easing numbers in the apartheid population equation. Typically of the apartheid state, the people involved were not consulted as to their wish or otherwise to become Swazi citizens. On the Swazi side there were two potential major prizes to be gained from the deal: a long-held irredentist dream, particularly on the part of the octogenarian king, Sobhuza II, and access to the sea. Agreeing to banish all ANC personnel, except a token diplomatic representation, did not bother the Swazi government, as it reduced the likelihood of cross-border raids by the South Africans either in hot pursuit of, or in retaliation for, ANC activity.

Swaziland, anxious to be seen to be behaving properly in the wider African context, cleared the deal with the then Organization of African Unity President, Daniel arap Moi of Kenya. The South African government, not used to having its proposals challenged internally, found itself in just that position. The puppet homeland government of KaNgwane objected to the deal, but their opposition carried little weight and was ignored. A more serious protest came from the homeland government of KwaZulu, led by Mangosuthu Buthelezi. Their concern was about Ngwavuma, the largely non-Swazi territory between Swaziland and the sea. The people living there were also largely not Zulu, and historically Ngwavuma had never been part of Zululand. But in 1976 the South African government 'gave' this isolated piece of land to the homeland of KwaZulu, again without any

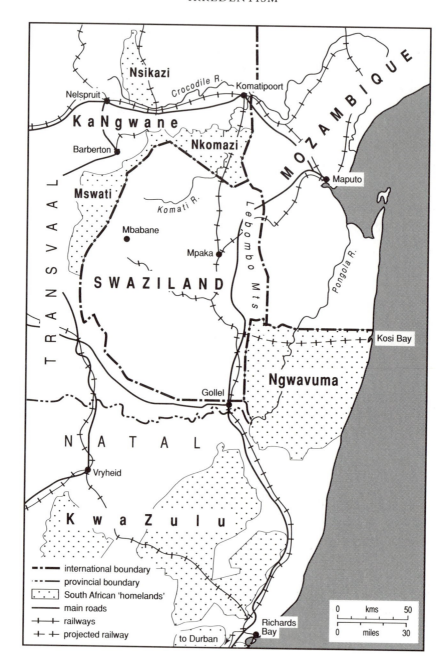

*Map 24* Swaziland

146

consultation of the inhabitants. To do so fitted in with the aims of white South Africa to consolidate and enlarge homelands to make them seem more respectable and more viable political units to the outside world at minimum cost. Buthelezi challenged the South African government in the courts, in effect claiming that the same piece of territory could not be given away twice to different people. KwaZulu also stressed the tenuous case Swaziland had in claiming Ngwavuma, whilst trying to exaggerate their own case which, in fact, apart from Ngwavuma being a recent gift, was even thinner. When the Natal provincial court ruled in favour of the KwaZulu homeland government the South African government took the case to the national Appeal Court which upheld the Natal court's ruling. Short of changing the law, Ngwavuma had to be removed from the land deal, and with it went Swaziland's hope of access to the sea.

As a result the proposed land deal was dropped. For the South African government, which was quite capable of retrospective changes in the law, the significance of Swaziland was reduced by the singing of the Nkomati Accord with Mozambique. For Swaziland a deal without Ngwavuma and access to the sea was not worthwhile, especially as the old king, Sobhuza, the enthusiast for the irredentist dream, had died. On its own the KaNgwane homeland was no great prize, with few resources and a very high density of population. The large numbers of people that Swaziland would have had to absorb represented a considerable political risk. They would have brought little in the way of resources to have enhanced the Swazi state and many of them did not want to become part of Swaziland anyway. Swazi political stability would have been put at risk and especially the fortune of the ruling elite round the Royal House, which had been plunged into some disarray with local rivalries being played out after the death of the old king. In different circumstances, if the timing had been different, the outcome might also have been different. Perhaps Africa would have one less land-locked state, and one example where an irredentist dream had been realized, with a Swazi nation-state more than twice its present size in both area and population.

That irredentism can be a potent political force is without doubt. That states are often willing to make considerable economic sacrifices to attain the political, emotional, even romantic, ideal of irredentism is certain. The paucity of examples in Africa is mainly to do with the number, size and distribution of ethno-linguistic groups in relation to the modern states of Africa. The partition of Africa paid scant attention to pre-existing culture groups and it is no thanks to the European powers who carved Africa up between them that irredentism is not a more widespread problem. Irredentism does exist in Africa and in the case of Somalia is at the heart of a serious and seemingly intractable problem.

# 13

# AFRICAN IMPERIALISM

Africa is the continent which, more than any other, has been dominated by imperialism. That some of the imperialism suffered by Africa and Africans is essentially home-grown is not really surprising. In some parts of Africa the political skills and organization necessary to create empires were developed long before European colonialism engulfed the continent. African empires such as Ethiopia and Morocco long survived against the European powers despite their technological inferiority and, succumbing briefly, have re-emerged with their imperialistic ideas intact and their imperialistic abilities enhanced from their close contact with the European powers and latterly the world super-powers. African imperialism has been accepted by the outside powers and even encouraged by them. African imperial powers have contributed greatly to the discomfiture and deep suffering of their neighbours, for theirs is a land-based imperialism spreading through contiguous territories. Their predatory existence in post-colonial Africa is another part of the African inheritance.

The Christian empire of the Prester John was known to Europe before European navigators rounded the Cape and opened up the Cape sea route to India at the end of the fifteenth century. The Prester John had sent representations to his fellow Christian King of Portugal who regarded him as an equal and an ally in the fight against Islam. Also an ally well placed for a potentially lethal strategic strike at the Islamic underbelly. Such military fantasies did not materialize but Abyssinia (Ethiopia) was always treated as an imperial equal, an ally to be assisted, a normative political entity.

During the nineteenth-century European partition of Africa, Ethiopia was again regarded by the European powers as their equal in imperialism. The aggressive imperialism of the Emperor Menelik in extending his territory at the expense of his neighbours in the 1880s and 1890s was accepted by the European powers and it was not until the Italians belatedly entered the 'scramble', in desperation for an empire, that Abyssinia was really threatened. In 1896, at the battle of Adowa, the Italians were comprehensively defeated and retired to their Red Sea colony to lick their wounds and nurture their grandiose imperial ambitions for forty years. Italian fascism in the 1920s and 1930s yearned for an empire to demonstrate its political virility. With the appeasing connivance of Britain and France, who with the infamous Hoare-Laval pact opened the Suez Canal to the Italian imperial force, Abyssinia was targeted. Deploying modern armaments, the Italians quickly won the war, and the short-lived Italian East African Empire created in 1936. Abyssinia's colonial experience in Italian hands was for five years only before being liberated by British and Commonwealth (South African, Nigerian and East African) forces who entered Addis Ababa after a lightning advance on 6 April 1941. Haile Selassie was back on his throne five years to the day after having to flee his capital before the Italian invasion.

Defeat by imperialism and exile had not spoiled Haile Selassie's imperialist appetite. Ethiopian imperialism went on to outlast direct European imperialism in Africa, and despite dramatic changes in its domestic political ideology Ethiopia was until 1991 a loosely-knit empire which had not succumbed to the forces of disintegration which had devastated much grander empires elsewhere. The main victim of post-war Ethiopian imperialism was Eritrea, but other parts of the empire, including Tigray in the north and the Ogaden in the east, have fought militarily for greater autonomy if not outright secession.

Eritrea contains within its colonially-drawn boundaries a wide diversity of landscapes, peoples and cultures. It comprises a narrow strip of land along the Red Sea coast, over 600 miles (1,000 km) long, which widens in the north to include a high plateau extension of the Ethiopian highlands and beyond that a western lowland bordering on the Sudan. Tigrinya speakers, who live on the plateau, make up about half the population of Eritrea and are mainly Christian who share their language and culture with their neighbours in the Tigray province of Ethiopia. Tigrinya speakers of the western lowland and the northern coastal strip make up about one-third of the population and are Muslim. In the southern coastal strip the Danakil are Muslim nomadic herdsmen related to the Afar of neighbouring Djibouti. Other small language groups in the north are mainly Muslim with Christian minorities. In general the coastal and western lowlands are inhabited by Muslims, the plateau by Christians.

Eritrea knew no unity before Italian colonization. From the sixteenth century the western lowland and northern coastal strip were part of the Ottoman Empire, which extended along the Red Sea littoral. The Turks were succeeded in the nineteenth century by Egypt and in turn by the Mahdist state. Ethiopia held the allegiance of the plateau area but, until the nineteenth century, showed little interest in the Red Sea coast. Before that, for most of recent history, Abyssinia was a land-locked Christian empire dependent on its highlands for isolated survival as an island in a sea infested by the hostile surrounding forces of Islam.

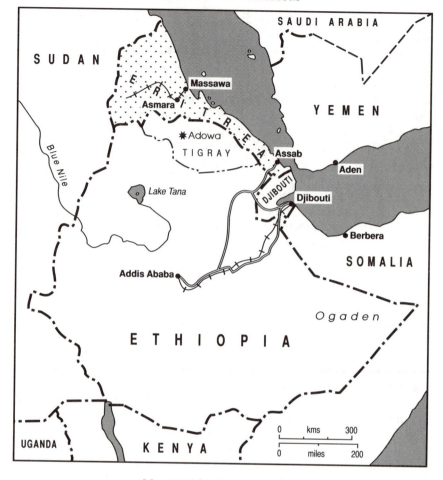

*Map 25* Ethiopian imperialism

During the European scramble for Africa the ports of Assab and Massawa became Italian colonies in 1882 and 1885 respectively, and in 1890 they were incorporated into the newly-formed Italian colony of Eritrea, which included the whole of the coastal strip between British Sudan and French Somaliland. The boundaries of the new colony were, as usual, drawn by the Europeans. Even the name Eritrea (Erythrea) was derived from the classical name for the Red Sea. The Italian attempt to declare a 'protectorate' over Ethiopia failed and they concentrated their colonial zeal on Eritrea, which in some respects became a model colony with fine roads and colonial buildings. For the first time, Eritrea was welded into a single political entity with unified political and social structures which cut across the traditional divisions. It broadly followed the pattern of political

150

development experienced in all other European colonies in Africa and which, in the vast majority of cases, formed the basis for eventual independence.

Between 1936 and 1941 Eritrea, along with Italian Somaliland as part of the Italian East African Empire, was ruled together with Ethiopia for the first time. In 1941, after the Italians were defeated, Eritrea, Somaliland and the Ogaden were placed under British Military Administration. Ethiopia, apart from the Ogaden, regained its independence under Haile Selassie. After the war Eritrea's future status was to be decided, like that of the other Italian colonies, Somaliland and Libya, but not Ethiopia, by a Four Power Commission of Britain, France, the Soviet Union and the United States.

The four powers could not agree on Eritrea's future, arriving at four different proposals: union with Ethiopia (Britain), partition with the highlands and southern coastal strip going to Ethiopia (United States), Trust Territory with Italian administration (France) and Trust Territory with international administration (Soviet Union). The problem passed to the UN who set up a Commission of Burma, Guatemala, Norway, Pakistan and South Africa, which was also divided. Partition was rejected outright. Guatemala and Pakistan proposed the standard formula of UN Trusteeship leading to independence, but the majority favoured close association with Ethiopia. Burma and South Africa favoured federation with some autonomy, Norway wanted full union. The United States backed federation with Ethiopia, and with only nine votes against (including that of the Soviet Union) UN Resolution 390A of December 1950 was passed. From September 1951 Eritrea became an autonomous territory federated with Ethiopia. The preamble to the resolution referred to Ethiopian claims on Eritrea: 'based on geographical, historical, ethnic or economic reasons, including in particular Ethiopia's legitimate need for adequate access to the sea'.

It also expressed a desire: 'to assure the inhabitants of Eritrea the fullest respect and safeguards for their institutions, traditions, religions and languages as well as the widest possible measure of self-government.'

How empty these words were, as was soon demonstrated.

Within Eritrea there emerged a Unionist party based in the highlands and an 'Independence Bloc' of parties broadly favouring independence. Ethiopia, allowed great latitude to influence affairs in Eritrea by Britain, financed the Unionists and intimidated the Independence Bloc with a terrorist campaign against its leaders and supporters. What was decisive was the alliance between the United States and Ethiopia who concluded a joint Defence Pact in 1953. British sources in Eritrea at the time were of the opinion that a majority of Eritreans would have voted for independence, but they were never given the opportunity of expressing their views in an official referendum of self-determination.

Ethiopia consistently abused the terms of the UN Resolution and systematically set about turning federation into full union. Amharic became the official language and the 'autonmous' government was blatantly interfered with. Elections were held without UN supervision and a puppet regime installed to vote for union with Ethiopia. The absorption of Eritrea excited little outside interest as the matter was considered internal to Ethiopia. Not for the last time in Africa, let alone elsewhere, this principle of non-interference in the affairs of an independent state was to allow a central government to assert its will over the people of a dissident region. At this time Ethiopia commanded considerable prestige and the feudal emperor's autocratic style impressed in international affairs. Haile Selassie became the father-figure of the first decade of African independence, an African who had triumphed over colonialism, whose pride and dignity had shamed the conniving politicians of pre-war Britain and France as well as the strutting Mussolini. Haile Selassie secured for Addis Ababa the headquarters of the UN Economic Commission for Africa (1958) and the OAU (1963) and with them implicit endorsement for his government and all its sometimes dubious works.

The war in Eritrea escalated into fully-fledged guerilla warfare on the one hand and massive retaliation by conventional land and air forces on the other. Almost inevitably the Eritreans divided, the more radical Eritrean People's Liberation Front (EPLF) challenging the Eritrean Liberation Front (ELF) and both indulging in internecine warfare. The Eritreans in general were portrayed in the West as left-wing Muslim dissidents attacking conservative Christian Ethiopia. Their action threatened to undermine United States strategy for the whole Middle East, which centred on the survival of Israel, by attacking the Red Sea naval bases which helped keep the southern sea route to Israel open. When the Ethiopian revolution of 1974 overthrew Haile Selassie, a neo-Marxist military government was installed in his place. The United States was dismissed as chief patron and the new Ethiopian government turned to the Soviet Union. With regard to Eritrea the new government was every bit as imperialistic as the old Emperor and the situation remained essentially the same. By the end of 1977 the Eritreans had gained control of all the territory except for some garrison towns but instead of negotiating, the Ethiopian Mengistu regime, now backed by the Soviet Union and Cuba, sought a military solution. In 1978 a newly-equipped Ethiopian army of over 100,000, with Cuban and Soviet support, was launched and retook almost all Eritrea at considerable cost to itself and Eritrea. Thousands of Eritreans were killed and hundreds of thousands of refugees fled across the northern border into the Sudan. But despite pouring a vast amount of money into military equipment, including MiG jet aircraft, Ethiopia was unable to deliver a *coup de grâce*. The Eritreans clawed their way back into contention and a 'fluid stalemate' prevailed where Ethiopia was unable to eliminate Eritrean guerillas and the Eritreans were unable to control all their territory or take the key garrison towns which were constantly under seige.

Ethiopia's position was made worse by a revolt in Tigray province, not for independence as in the case of Eritrea, but for greater autonomy within Ethiopia. Eritrea and Tigray were devastated by the droughts of 1983–5; thousands died of starvation but the wars relentlessly went on. The human suffering was appalling but people outside were aware only of the tip of the iceberg. As the 1980s progressed the war took its toll on Ethiopia. The military dictatorship spent vast sums on the army and air force despite the desperate plight of millions of people through recurring famine. In addition, they attempted at the same time a collectivization of peasant agriculture and tried to resettle up to 1.5 million people to overcome the effects of the droughts in the north. These were inappropriate, imperialistic, ideological and dictatorial responses to the problems that faced Ethiopia and all were unsuccessful. In 1991 the Mengistu regime was at last defeated militarily by a force that fought its way out of Tigray to take an Addis Ababa half-heartedly defended by a demoralized army. Mengistu and his family flew to a prepared refuge in Zimbabwe whence he has yet to be extradited.

For thirty years from 1960 Eritreans fought for their independence from Ethiopia. Eritreans viewed the long-lasting conflict as a fight for the basic human right of self-determination denied them in the past by the UN. They regarded Eritrea as a separate political entity which was forced into federation and then union with Ethiopia. Ethiopia, on the other hand, regarded the conflict as a secessionist war waged by a rebellious region which, if successful, would have left Ethiopia land-locked and in danger of further disintegration. The two positions were irreconcilable. It took the fall of the Mengistu regime to the Tigrayan-led rebellion in 1991 for Ethiopia to stop the war against Eritrea. Peace between Addis Ababa and Eritrea, following the installation in Addis of a regime with regional roots who understood and sympathized with the Eritrean predicament led to a settlement. Eritrea was allowed to proceed to full independence in May 1993 following a long-delayed referendum which, not surprisingly, gave overwhelming support for a status long and bitterly fought for.

The settlement, followed by the independence of Eritrea, has left Ethiopia a land-locked state once more. The colonial boundary between Eritrea and Ethiopia, defined in a treaty between Italy and Ethiopia in July 1900, became the international boundary between the two sovereign states without modification. There had been speculation that Eritrean independence would have been bought by ceding to Ethiopia the port of Assab with a corridor of land containing the paved road to Addis Ababa, but this did not materialize. Whether the permanence of the settlement will be affected by this time will tell, but access to the sea is a powerful force which may well bear on future Ethiopian governments.

The Western Sahara problem is essentially similar to that of Eritrea. On ceasing to be a European colony Spanish Sahara was occupied by a neighbouring African state, Morocco, which claimed historic pre-colonial rights. The territory, because of its natural resources, phosphates and off-shore fishing, rather than its strategic location, is also of considerable value to the occupying power. The people of Western Sahara have been denied the right of self-determination and the guerilla war fought by the Sahrawis for independence against a powerful militarized state with super-power backing has literally run into the sand. Unlike the situation in Eritrea the international community has been involved with the Western Sahara dispute at several levels: the International Court of Justice (ICJ), the United Nations (UN) and the Organization of African Unity (OAU). The net effect is painfully slow progress towards a referendum to determine the future of the territory, but that could well solve nothing as accusations are made that among those eligible to vote will be large numbers of Moroccans who have moved southwards across the border into Western Sahara since 1975.

In 1884, at the height of the European scramble for Africa, Spain claimed the 600-mile (960km) Saharan coast between Morocco and Mauritania as its 'sphere of influence'. Spanish enthusiasm was limited to the foggy hot desert coast washed by the cold Canaries Current and awash with rich stocks of fish. The Saharan coast also faced their valued possession of the Canary Islands, now ruled directly as a province of Spain, and so there was a strategic element in the Spanish acqusition of the Saharan coast.

France became politically dominant in Morocco from 1911, after imperialistic deals with Great Britain and Germany. However, Spain secured those parts strategically important to it: the northern Rif, facing Spain itself and the hinterland to Spanish ports of Ceuta and Melilla on the north Moroccan coast; the southern protectorate of Morocco opposite the Canary Islands and contiguous with Spanish (Western) Sahara; and the fishing port enclave of Sidi Ifni on the Atlantic coast of southern Morocco.

Spanish interest in and control over the desert interior was minimal. The population was small and beyond the few coastal settlements comprised sparsely distributed nomadic herders whose traditional territories extended beyond the mainly straight-line boundaries drawn by the European powers. The Spanish census of 1974 put the total population of Western Sahara at 73,500 but the UN estimated more than twice that number. At the higher figure the density of population was one person for almost two square kilometres. In 1965 deposits of an estimated 1,700 million tonnes of high quality phosphates were confirmed at Bu Craa. A refinery was built near the open-cast mine and connected by trunk conveyor belt to the port of El Aaiun (Laayoune). At a stroke the northern part of Spanish Sahara was transformed from a desert wasteland into valuable real estate.

On attaining independence for the bulk of its territory from the French and Spanish in 1956 Morocco was aggressively expansionist. It claimed

Spanish Sahara, French-ruled Mauritania and parts of Algeria on the basis of the sixteenth-century Moroccan empire which had extended as far as Timbuctoo at the southern edge of Sahara in French Soudan, soon to become Mali. In 1957 Morocco invaded Spanish Sahara in pursuit of these claims but was repulsed by Spain. Following Algerian independence was flared between Morocco and Algeria in 1963. The conflict centred on the large unworked iron-ore deposits at Tindouf and those parts of the international boundary which the French had failed to define when administrating both territories. Moroccan aggression was again repulsed.

Diplomacy took over: Spain gave up the Sidi Ifni enclave in 1968 but left the issue of Spanish Sahara, Ceuta and Melilla to be resolved. Morocco and Algeria signed a treaty of friendship in 1969, and in the same year Morocco recognized Mauritania, relinquishing its former territorial claims on that state. Morocco continued to press for decolonization of Spanish Sahara with the assumption that Spain's withdrawal would be followed by Moroccan rule. But Spain's belated phosphate-led interest in the Sahara resulted in some economic development and an accompanying political awakening. In 1967 Spain set up the Yema'a, an assembly of nominated and elected Sahrawi members to give advice to the Spanish colonial government on local administration. In 1973 the Yema'a asked that the Sahrawis be accorded the right to self-determination on the basis of the colonial boundaries and, in 1974, to Morocco's consternation, Spain agreed. Meanwhile, in May 1973, a new nationalist movement, the POLISARIO front (*Frente Popular para la Liberación de Saguia el Hamra y Río de Oro*) had been formed to accelerate political development by direct action and win independence for Western Sahara.

The advent of the POLISARIO galvanized Morocco into diplomatic action. The UN was persuaded to ask the ICJ to advise on the legal status of Western Sahara before Spanish colonization. The UN also agreed to send a mission to assess the problem on the spot and to visit other interested states. Spain agreed to postpone a referendum in Western Sahara until the UN had received both reports. Late in 1974 Morocco and Mauritania secretly agreed to partition Spanish Sahara between them when the opportunity arose.

In mid-October 1975 things came to a head dramatically when, within a few days of each other, the UN mission and the ICJ published their separate findings as Franco, the Spanish dictator, lay dying. Both reports recognized pre-colonial ties between Western Sahara and Morocco and between Western Sahara and Mauritania but, on the other hand, saw no reason to withhold from the Sahrawis the right of self-determination. Spain, playing for time in an awkward interregnum, went back to the UN, who suggested a six-month cooling-off period. King Hassan seized the opportunity presented by the Spanish indecision and on 6 November 1975 led 350,000 Moroccans in the well-publicized 'Green March' across the Saharan border. On 14 November an agreement was signed between Spain, Morocco

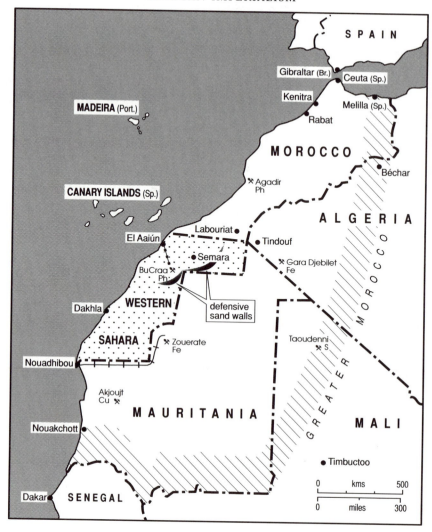

*Map 26* Moroccan imperialism

and Mauritania for Spanish withdrawal in early 1976 and for the partition of Western Sahara between Morocco and Mauritania, Morocco taking the northern two-thirds with the phosphates of Bu Craa.

Excluded from the agreement and ignored in the take-over, the POLISARIO fought on and, in February 1976, the Saharan Arab Democratic Republic (SADR) was set up. With help from Algeria, which gave effective sanctuary, the POLISARIO first chipped away at Mauritania, whose vital iron ore mine at Zouerate and railway to the port of Nouadhibou could not be easily defended against guerilla attack, even with French airforce help (1976–8). The unpopular war threatened Mauritania's main resource, drained its fragile economy, built up the army and led, almost inevitably, to a military *coup d'état*. The long-serving Ould Daddah government fell, a cease-fire with the POLISARIO followed and in 1979 Mauritania made peace with the SADR, renouncing the partition of Western Sahara agreed with Morocco.

The war between the POLISARIO and Morocco settled into stalemate. The POLISARIO controlled three-quarters of the territory of Western Sahara but Morocco held the part that mattered economically – the Bu Craa – El Aaiun – Semara triangle. The Moroccans pushed out walls of sand hundreds of miles long to form a line of defence effectively to protect the triangle. This well-prepared defence wall, topped by sophisticated radar devices supplied by the United States, which has a major military staging-post base at Kenitra on the northern Atlantic coast of Morocco, prevented the POLISARIO from continuing their earlier devastating surprise guerilla strikes from their Algerian safe haven bases. They were free to roam at will through the sandy wastes of much of the south or to hurl themselves against the defensive wall in costly, futile attacks. Behind the sand walls the Moroccans moved to integrate Western Sahara with Morocco, including the building of a tarred road south from Agadir, and have discovered large new deposits of iron ore which they are planning to exploit.

On the diplomatic front the POLISARIO/SADR steadily gained support and recognition within Africa, but that support had no immediate effect on the situation. A majority of member states were willing to admit the SADR to the OAU at the 1980 Sierra Leone summit, but there was prevarication. An OAU ministerial meeting at Addis Ababa in February 1982 admitted the SADR by a simple majority but Morocco and its supporters withdrew, claiming the meeting inquorate as less than two-thirds of member states were represented. The most serious among several incidents was that the OAU failed to meet at Tripoli in August 1982 because more than one-third of member states stayed away. The OAU was unable to solve the matter and openly split into conservative and radical camps, giving a stunning and damaging display of African disunity. In 1984 the SADR was seated at the OAU with support in particular from Ethiopia, Angola, Mozambique and Zimbabwe. Morocco responded by walking out. But the recognition has led nowhere. In leaving the OAU Morocco has not suffered any obvious damage to its position in Western Sahara.

The Western Sahara problem was stalemated when an apparently simple demand for self-determination on the one hand and imperialistic claims on the other got caught up in the tangled and interwoven webs of African, Arab and super-power politics. Militarily the POLISARIO could only irritate the entrenched Moroccans but they would not go away and, as long as they had a safe base in Algeria, they could not be defeated decisively.

Diplomatically the only way forward was through a referendum, as suggested by the OAU in 1981 and later taken up by the UN. But Morocco has consistently refused prior withdrawal and there is seemingly endless wrangling over the vital composition of the electoral roll.

Supporters of both sides are divided, not necessarily on the merits of the case but according to traditional loyalties, political prejudice and self-interest. Within Africa, Morocco's chief supporters, Senegal, Somalia, Tunisia and Zaire want the *status quo* to remain, fearing that a POLISARIO victory would strengthen radical forces within the OAU. Saudi Arabia, the United States and France share these concerns on wider fronts and are prepared to support Morocco with financial aid, military supplies and diplomatic influence. Morocco's Kenitra base has served the United States well as a military staging-post though it may now be coming towards the end of its usefulness. It is also well recognized that Western Sahara is an issue by which King Hassan is able to deflect criticism of his sometimes precarious and unpopular rule within Morocco. Better the conservative king closely allied with the West and the other traditional Muslim kingdoms than a more radical, perhaps fundamentalist, replacement regime.

Support for the SADR has primarily come from Algeria who have given the all-important sanctuary for the refugee camps, into which the POLISARIO melt away, on their territory. Algeria is concerned about Moroccan expansionism in general but in particular would like to develop the Tindouf iron ore mines with access to the sea via El Aaiun. In April 1983 Algeria and Morocco moved closer together and for the first time in almost eight years the long land boundary between them was opened. Libya has been at best a fitful SADR supporter and in late 1984 Gadafy recognized Morocco's case in Western Sahara in return for Morocco's recognition of the Libyan case in northern Chad: again the diplomatic support given was not demonstrably related to the merits of the Western Saharan case.

Morocco has invested heavily in its part of Western Sahara and is extracting minerals quickly. It is highly unlikely that Morocco will countenence a withdrawal even if there was diplomatic pressure so to do. Such pressure does not exist from the quarters where it might count. There are constant reports of a referendum being agreed and equally frequent reports of a referendum being postponed. The longer and wider Moroccan occupation of Western Sahara is, the more difficult it will be to hold a meaningful self-determination referendum.

During the process of decolonization in Africa the norm has been for every European colony to achieve independence as a separate political entity. In the few cases where two or more colonies came together to form a single state, for example, in Somalia, South Africa and Tanzania, it was on the basis of voluntary agreement of all parties. In Somalia and South Africa union coincided with independence with the co-operation of the colonial powers, in Tanzania it came shortly after the independence of Zanzibar and two-and-a-half years after that of Tanganyika. In their Trust Territories, including the partitioned Cameroons and Togo, the UN adhered to the principle of self-determination with referenda, even if grossly delayed in the case of Namibia and a hotly disputed overall result in British Togoland.

Eritrea and Western Sahara were both prevented from achieving independence mainly through the action of a contiguous imperialistic state. Ethiopia and Morocco attempted to ignore the period of European colonialism and to resume, as if uninterrupted, nineteenth-century imperialism. In this they were aided and abetted by the UN and the world super-powers. Where, in Eritrea, the UN had direct responsibility, the right of self-determination was denied the people of that former Italian colony largely through pressure exerted by the United States. In Western Sahara, the UN failed to prevent Morocco maintaining *de facto* occupation and procrastinating over a referendum whilst at the same time taking steps to ensure a result favourable to its *de jure* annexation of the country if and when it should take place. Again the political presence of the United States has been evident. In both cases prolonged guerilla wars ensued. Military action in support of self-determination and independence in both cases has been tenacious. Neither Ethiopia nor Morocco could have maintained their struggles to forcibly annex their neighbours without the diplomatic, material and military support of outside powers, including the United States and the Soviet Union, with their own particular priorities. The merits and demerits of the opposing causes have generally had little to do with the outside support given. In the case of Eritrea the UN and the OAU ignored the war on the grounds that it was a civil war, a secessionist war, a matter internal to the state which should not be interfered with. In Western Sahara the UN and the OAU have intervened only to have their efforts effectively ignored. This demonstration of impotence long pre-dated Bosnia and the new world order. Neither problem has been solved directly by the war it created. Eritrea has achieved independence because it fought a tenacious thirty-year-war, because the military regime it was fighting was overthrown by a different force, a third party, and because of the ending of the Cold War terminated super-power interest in the region.

African imperialism has been part of the post-colonial African inheritance, tolerated because it has been perpetrated by African powers who were useful to the wider aims of the Cold War super-powers, and in the case of Morocco in the 1990s, useful in countering the spread of Islamic fundamentalism. Outside forces have been willing to supply almost unlimited arms and other military assistance in pursuit of their own perceived strategic advantage and economic gain. The ending of the Cold War can be said to have contributed indirectly to the emergence of Eritrea as an independent state but the Western Sahara problem drags wearily on.

# 14

# POLITICAL UNION

Antidotes to the effects of the African inheritance are, in theory, many, but in practice are difficult to administer. The history of post-independence Africa is strewn with attempts to put in place structures designed to ameliorate the condition. Some African leaders forecast what was to happen and advocated preventive action, but that advice was in general ignored in the stampede for the independence on offer from the colonial powers. The goal of independence appeared to be such an improvement on the dependent colonial state that few even bothered to try to look beyond it. Independence promised just that, independence in political and economic spheres and, within the constraints of the international order, control of one's own destiny. The few, often regarded as prophets of doom denying the light just as the dawn of independence was breaking, saw that the structures about to be inherited in themselves condemned Africa to a subservient position in world affairs and clearly foresaw the pre-conditions for blatant neo-colonialism. Most of the European colonies in Africa were too small to become effective independent states. They would be too weak to change Africa's lowly place in the world economic system. They would likely continue to produce from their finite natural resources raw materials for the consumption and use of their colonial masters, though of course the names would be changed to disguise the barely changed relationship. The trappings of independence and new-found national pride along with the very real rewards for very few were dazzling lights which blinded almost all to the grave dangers which lay ahead.

One of the few African leaders to see clearly the dangers ahead was Kwame Nkrumah, the first President of Ghana, the first black African state to win its independence in March 1957. He articulated his views in a book, *Africa must unite!*, which largely went unread, advocated them at meetings with fellow heads of state, but was not heard. His further prophetic but also unheeded words pointed to: 'the necessity to guard against neo-colonialism and balkanization, both of which would impede unity' (Nkrumah 1963: 138). In the more than thirty years since those words were penned the first generation of post-colonial Africa has witnessed little

160

unity but has experienced the severe deleterious effects of neo-colonialism and balkanization.

African independence was generally achieved on the basis of small political units. The majority of African states are small in area, in population, in size of market and economy, in terms of political clout. A few African states, for example Zaire, Nigeria and Sudan, are territorially so large, so populous and culturally so varied as to suffer from almost the opposite drawbacks, but the majority of African states are politically and economically weak largely because of their smallness as measured in one or more critical ways. European colonization certainly brought many small culture groups under single political administrations but it also divided Africa between the various European powers, each with its own political and administrative traditions and language, so that for example in West Africa, no British, German, Portuguese or Spanish colony was contiguous with another colony of the same European power. The French in West Africa, the British in southern and East Africa and briefly the Italians in East Africa held contiguous tracts which were sub-divided into individual colonies. Where they held contiguous colonies the colonial powers felt the need to federate the individual colonies into larger political units the better to enjoy economies of scale in aspects of administration and development. Thus the French created two great colonial federations (*Afrique Occidentale Française* and *Afrique Equatoriale Française*) in 1902 and 1908 respectively, the Italians established their short-lived *Impero dell'Africa Orientale* (1936–41), and the British later (1953–63) put together, again briefly, the Federation of Rhodesia and Nyasaland. Although the colonial powers obviously saw advantages to be gained from these larger administrative units in Africa they did not feel the need to pass them on when granting independence. On the other hand, from the post-independence experience of Zaire, Nigeria and Sudan, there were dangers in creating large states in Africa without an over-arching outside power to impose unity in a way that could not be achieved with self-government. The European empires in Africa, in this respect, fulfilled roles similar to that of the Soviet Union and Yugoslavia in supressing ethnic or regional rivalries which came to the surface again when the wider, binding authority was removed.

None of the large colonial federations survived the transition to independence. It would have been surprising if either of the large French federations survived intact because of their enormous size, but in the event not even two or three colonies stayed together as a single state. Independence was achieved on the basis of the individual component colonies. Reasons for the balkanization of the contiguous European empires varied. In the AOF, apart from its vast size, rivalry between the African leaders of individual colonies, as between Leopold Senghor of Senegal and Felix Houphouet–Boigny of Ivory Coast made any large-scale union of colonies difficult. The opportunity to achieve independence on the basis of the individual colony induced many African politicians to take that option, with the certainty of a small gravy-train, rather than run the risk of losing out to a rival in the competition for a larger fiefdom. Italian East Africa was broken up following the military defeat of Italian forces there during the Second World War, though the Ethiopian take-over of Eritrea strictly represented a union of two of the three former colonies that made up that empire, albeit a forced and, after thirty years it can be said, temporary union. The Federation of Rhodesia and Nyasaland was created largely to satisfy white settler interests. It broke up because the tidal wave of African independence could not be prevented from engulfing the two northern territories but could be halted, at least temporarily, at the Zambesi. Settler domination of Southern Rhodesia effectively stopped the Federation from achieving independence as a single state as black majority rule was delayed there from 1964, when the northern territories became independent, until 1980, when Zimbabwe achieved its independence. That is not to say that the Federation had sufficient political and economic rationale to have survived independence without the settler factor which had created it in the first place. Whatever the cause, balkanization, the division into small component political units, certainly made the task of economic and political development harder to achieve and neo-colonialism easier to impose. To what extent the balkanization of Africa was a deliberate policy on the part of the departing European powers is debatable. They did little to prevent it, or to encourage unions of colonies, but on the other hand balkanization was given most support by African leaders who sought to benefit from it.

Territorial units which made acceptable individual colonies or protectorates did not necessarily make effective independent states. This is the other side of the coin of removing the over-arching authority and rationale of empire. As independent states many former colonies were often too small, too sparse in population, too poor in resources, too weak politically and economically and too remote. Most of the fifteen land-locked states of Africa, once the distant parts of contiguous and sometimes federated empires, are like this. There was a need to create larger units, not to have more Zaires or Sudans, but to have fewer Burkinas or Burundis.

The obvious time to create such states was at the point of independence. Two or more territories, part of the same colonial empire, but with no history as independent states could have been more easily welded together as a single political entity than could established states each with its own traditions, governments and administrations. This was the argument of the pan-Africanists whose cause was espoused within Africa by Kwame Nkrumah. Political union was a necessary prerequisite to economic union. That view was endorsed by the Casablanca bloc, the opposing view by the Monrovia bloc. Both groups were subsumed by the Organization of African Unity in 1963 where the views of the latter bloc gained the ascendancy.

Very few African states are unions of colonies. The exceptions are South Africa, a union of four former British colonies, Somalia, the union of a former Italian and a former British colony, and Tanzania, a union of two former British colonies. British Togoland, a League of Nations and then a United Nations Trust Territory, was united with Ghana in 1961, the northern section of British Cameroons, another Trust Territory, was united with Nigeria and southern British Cameroons with Cameroon. Senegal has featured in two abortive attempts to unite with contiguous territories, with Mali in 1960 and with Gambia in 1982–89. Ghana, Guinea and Mali attempted union in the period 1958–63, despite Ghana being non-contiguous. Colonel Gadafy of Libya proposed several unions, with Egypt (1972), Tunisia (1974) and Chad (1982) but none came to fruition. The record of unions of colonies and states in Africa is therefore not impressive, though a more detailed examination of the experience is instructive.

The Union of South Africa was formed from the four former British colonies of the Cape, Natal, Orange River and Transvaal in 1910. Its birth was long and painful. In 1859 it was Sir George Grey, the British Governor of the Cape Colony, who at the opening of the Cape Parliament first expressed the idea of some form of federal union for South Africa, as he put it 'to confer a lasting benefit upon Great Britain' (Walton 1912: 7). The British attempt to achieve union by annexation of the Transvaal in 1878 ended in the first Boer War of 1881 which re-established the Boer South African Republic. It took the second Boer war of 1899–1902 to bring South Africa to the point where union could again be thought of. In 1902 the two former Boer republics of the Orange Free State and the Transvaal became British colonies, thus bringing the whole of South Africa into the British empire. Some, including the British High Commissioner in South Africa, Lord Milner, advocated that union should be pushed through as an extension of the post-war settlement. A change of government in Britain led to the two former Boer republics being given self-government to bring them into line with the two older British colonies of the Cape and Natal. It was from this position of constitutional equality that the four colonies came together, with the blessing of the imperial power, to thrash out the means whereby they could by united in a single sovereign state under the British crown. Union was preceded by a National Convention where delegates from each of the four colonies met in secret session to thrash out an agreement on a Union constitution. Once this was accomplished South Africa proceeded to Union in 1910 with the full blessing of the British Parliament at Westminster.

The issues that drove the four colonies to the negotiating table were economic, closely linked and long-running: railways and the customs union. 'The National Convention . . . was the result of the failure of the delegates from the several Colonies to come to any agreement on Customs Tariffs and Railway Rates' (Walton 1912: 26). Once there they had to deal not only with

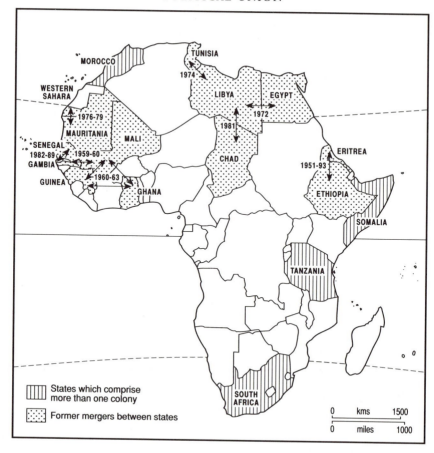

*Map 27* Post-colonial unions of states

those issues but also with constitutional issues: federation or union, how much power to the individual provinces, admission of other territories to the union; the franchise question, votes for non-whites; the language question, English and Afrikaans as official languages; the judiciary, the supreme court and appeal court; finance, public debt and taxation; the capital city, which of the four colonial capitals; the civil service; and this being South Africa, 'native' affairs and 'native' territories. This listing of only the main headings of the discussions of the National Convention indicate the complexity of attempting to attain a union of colonies. On 20 September 1909 the British Parliament passed the South Africa Act 1909 and the Union of South Africa came into being on 31 May 1910. Fifty-one years to the day later it became the Republic of South Africa, having achieved full autonomy from Britain along with the other 'white dominions' in 1926.

164

South Africa became the continent's strongest economic power based on minerals but with sufficient size to support the largest manufacturing sector of any economy in Africa. Politically the country has been dominated by the racial question, which featured large in the National Convention of 1909 and was the crux of the multi-party negotiations leading up to the 1994 elections and majority rule. It could be argued that in tolerating the extreme and uncompromising views of the former Boer republics over the non-white franchise, and leaving the Cape franchise eventually vulnerable to change, the 1909 National Convention sowed the seeds for the disaster of apartheid. It was a heavy price to pay for union but one that was not anticipated at the time. Natal was the most reluctant of the four colonies for union in 1909 and argued for a looser federation, but lost the day. In the 1990s Natal was again uneasy during the constitutional wrangling leading up to majority rule. Much of that uneasiness came from Mangosuthu Buthelezi and the Inkhata Freedom Party based in KwaZulu in Natal Province. Buthelezi received support from whites in Natal, notably for his *Indaba* in April 1986 to advocate the setting up of a multi-racial government in Natal, which smacked of the possibility of secession from the Republic of South Africa.

British Somaliland became independent on 26 June 1960 and on the 1 July joined with former Italian Somaliland to become the Republic of Somalia. The two colonies had been ruled as one briefly during the Italian occupation of British Somaliland in 1940–1, and the British military administration of Italian Somaliland in 1941–9. The *raison d'être* for the union was the ethno-linguistic unity of the two colonies as well as their economic homogeneity. The colonies were also extremely poor and very sparsely populated. British Somaliland had a population in 1960 of about 650,000 and Italian Somaliland about 1,250,000. The union still left about 1 million Somalis outside the rule of the Republic, in Ethiopia, Kenya and French Somaliland, leading to the irredentist wars which have done so much to further impoverish Somalia (see Chapter 12 above). In 1991, as civil war engulfed Somalia again, the former British Somaliland announced its secession from the Somali Republic. The old colonial boundary re-emerged as a fault line in the war-torn country. The secession has not been recognized by any other state or by the United Nations but Somaliland continues to proclaim its independence.

No other European colonies in Africa were joined together at independence. The Trust Territories of Togoland and Cameroons, administered by Britain, were never separate colonies but were ruled by the British during their Trust (later colonial) period as parts of Ghana and Nigeria respectively. Following UN referenda of self-determination in the Trust Territories, British Togoland was ceded controversially to Ghana and British Cameroons was divided between Nigeria and Cameroon.

Only in two instances have former colonies joined together subsequent to achieving independence and becoming separate states. In one of those cases the union continues and is almost thirty years old; in the other it lasted some seven years only. In December 1961 the ex-German colony and British-administered Trust Territory of Tanganyika achieved independence. It was the first country to achieve this status in East Africa, but was quickly followed by Uganda (1962) and Kenya (1963). The fourth British East African territory was the protectorate of Zanzibar, the seat of a non-European colonial empire, which had included, in addition to the islands of Zanzibar and Pemba, the east African littoral between the Rovuma and Juba rivers. The mainland territories had been ceded to the British and the Germans at the end of the nineteenth century to become part of Kenya (British East Africa Protectorate) and Tanganyika (German East Africa) respectively but the islands remained separate. They were ruled by the Sultan of Zanzibar, whose family had moved from Oman in the Persian Gulf in the early nineteenth century to take over the ancient Zenj empire. The empire was based on an Indian Ocean trade in ivory and slaves between the Zanzibar entrepôt and the Gulf. Having taken the Sultan's continental possessions the British and Germans struck a deal whereby Britain gave Germany the North Sea island of Heligoland in return for Zanzibar, which was made a British protectorate in 1890.

Zanzibar was given its independence on 10 December 1963, two days before Kenya. Within a month the Sultan's government was overthrown in a *coup d'état* led by the radical Afro–Shirazi Party. Three months later, on 27 April 1964, Zanzibar was led into union with Tanganyika by Abeid Karume, who had emerged as the leader of the Afro–Shirazi Party and President of Zanzibar, to form the United Republic of Tanzania. The union has been fraught with tension, despite Zanzibar retaining a large degree of autonomy. Zanzibar has maintained a post-independence tradition of volatile politics and has been less than happy with its share of development aid compared with mainland Tanzania. The union has survived but for long periods has looked particularly fragile. The initiative for the union came from Karume as a means of preserving his revolution in Zanzibar. His tyrannical rule, during which he kept Zanzibar at arm's length from the rest of Tanzania, ended in assassination in 1972.

The only union of colonies to emerge from French West Africa (AOF) was that between Senegal and land-locked Mali, which came together in 1958 towards the end of direct French rule. The fragile unity survived independence by only two months in 1960 before it broke up in acrimonious dispute, a closed border and a ripped-up railway line. Mali's traditional access to the sea was blocked and new routes were sought through Guinea and Ivory Coast. Mali then sought another abortive union with its other seaboard neighbour, Guinea.

In 1958 Guinea was the only French colony to reject General de Gaulle's offer of a future in the French Community. De Gaulle's reaction to the rejection was to grant Guinea immediate independence and to cut off all French assistance and withdraw all official French personnel. The calamitous results for Guinea of this fit of Gallic pique led directly to the first post-independence union of African states. Kwame Nkrumah in an expression of solidarity offered to unite Ghana (independent in March 1957) with Guinea, despite their lack of contiguity. The move was also consistent with Nkrumah's long-term aim of setting up a 'United States of Africa'. That positon was not yet lost and the battles between the Casablanca and Monrovia blocs were still in the future. After its failed union with Senegal, Mali also joined the Union of African States. Nkrumah saw the need for unity very clearly and hoped the 'Union of African States': 'would prove to be the successful pilot scheme to lead eventually to full continental unity' ... [other states, he said, were too] 'jealous of their sovereignty and tended to exaggerate their separatism' (Nkrumah 1963: 143, 148). The Union nominally lasted until 1963 but little progress was made towards a meaningful integration of the governments and services. It was no more than a union in name. However admirable, well-intentioned emotional responses are not an effective basis for unions of states which, in addition to good-will and trust between all parties, require much hard-headed negotiation, cool consideration and enormous attention to detail.

For the same reasons the various impulsive proclamations of unity between Libya and her neighbours have led to nothing except closed borders. Colonel Gadafy has gone to the extreme of announcing union with Tunisia without first even informing the Tunisian government. The various proposed unions have never progressed far beyond the announcement, showing on Gadafy's part little appreciation of the complexities of actually achieving a union of two sovereign states, though his initial motives might have been admirable.

In contrast to the above, the Federation of Senegambia survived for over seven years (1982–89). It was not the product of emotional response or a flush of rhetoric but of a long gestation, even if the actual birth was hurried and induced. The Gambia is Africa's smallest continental state, barely viable in political or economic terms, with a population of under 1 million and one of the lowest Gross National Products per capita in Africa. It is an enclave entirely surrounded by Senegal except for a short seacoast, with an almost totally artificial boundary comprising arcs of circles drawn from Gambia River and parallels of latitude. These geometric lines cut across culture groups so that the two countries are not distinctive from each other in ethno-linguistic terms. They do, however, suffer from having had different colonial experiences and have different official languages, French and English. Despite the Gambia being one of Africa's mini-states its config-uration, long and narrow, causes much disruption to Senegal. It has been described as a sword plunged into Senegal, as it makes Senegal a U-shaped territory with Dakar remote from the southern Casamance region which as a result has developed secessionist tendencies. From the Gambia's independence in 1965 the two governments worked closely together. The Trans–Gambian highway was built as a direct route between Dakar and the Casamance region and the Gambia showed its enthusiasm for co-operation by changing its traffic laws to conform with those of Senegal by directing traffic to drive on the left-hand side of the road. Political stability in both countries fostered co-operation, and the leaders Leopold Senghor and Sir Dawda Jawara came to know and trust each other. In 1973 on a visit to Dakar Jawara pronounced union with Senegal to be an inevitable and necessary development. Relations between the two countries warmed as successive schemes of co-operation were developed and points of friction were eliminated. The Gambia River Development Organization was born and in the mid-1970s the two countries sorted out the boundary ambiguities inherited from the British and French. Things came to a head suddenly. In July 1981 Jawara visited London to attend a Royal wedding. In his absence his political opponents staged a *coup d'état*. Jawara flew to Dakar and, with the assistance of Senegal, fought back. Senegalese troops were employed to help restore Jawara and the Senegalese policed Banjul for some time. Jawara immediately set about establishing closer links with Senegal as a means of safeguarding his position. On 1 January 1982 the Confederation of Senegambia was formally declared. The work of integration went slowly ahead: a confederal cabinet was formed and a parliament elected, both meeting for the first time in 1983. In the latter there were twenty Gambian members and forty Senegalese with decisions requiring a 75 per cent majority. Progress towards a fuller union ran into the sand as it became evident that the Gambia wished to retain its sovereignty whilst Senegal wanted full union. In September 1989 the confederation broke up to echoes of Nkrumah's prophetic words on individual states' sovereignty. Even the smallest state in Africa prized its independence and its political leaders clearly appreciated the perks of office (from which they were rudely ousted by military *coup d'état* in 1994).

Africa in the 1990s is divided into more sovereign states than ever before. Since the United Republic of Tanzania was formed almost thirty years ago, there has been only one serious attempted union of states, Senegambia, and that has not lasted. It is no longer a matter of colonies joining together, because there are no colonies left in Africa, but the much more difficult task of creating a union out of sovereign states. Nkrumah was right when he feared that if the opportunity afforded at the time of independence was passed up the task of unifying established states would be much more difficult. The South African experience of 1909 points to the complexity of the task of creating a union, but also shows that it can indeed be achieved, because if not completely independent the four South African colonies were self-governing and the imperial power did not interfere in the details of achieving a union. The Ethiopian experience with Eritrea points to the fact that union with an unwilling party might be forced but at enormous cost and to end in ultimate failure. Nevertheless Morocco persists in trying to force the people of Western Sahara into an unwanted union. Unity, which can be strength, and a strength much needed in Africa, must be between willing parties and based on agreed aims which take time to work out, and even longer to implement. In a careful survey of the continent there seems to be no real prospect of any African states achieving such a unity in the foreseeable future.

# 15

# ECONOMIC GROUPINGS

The need for African states to come together to form larger economic units is widely accepted. The arguments between the Casablanca and Monrovia blocs in the late 1950s and early 1960s were, crudely, about whether political union should become before economic union or *vice versa*. There was little doubt expressed then, or since, that African countries would be better served if they could co-operate closely together in economic matters.

The arguments in favour of economic co-operation are overwhelming, especially given the political balkanization of Africa. Most of the states of Africa are too small to be economically viable. Most African states are too small effectively to protect their resources from exploitation by large multi-national companies with their specialized expertise and budgets which far exceed the total Gross National Product of many African states. African states are too small to do anything to influence the world commodity markets on which they depend for their export income from raw materials. They are too weak to do anything effective to break a world economic system in which they occupy the lowliest of positions, exporting raw materials at low prices and importing manufactured goods at high prices. As national economic markets most are too small support anything other than the most small-scale, ubiquitous and primitive of industry. African economies are generally too small to generate sufficient income to thoroughly develop infrastructure within the state and between African states, something which must be done if Africa is to break out of its subservient global economic role.

The litany of 'too small . . .' goes on and on. It extends from the economy as such into closely related fields like education. Countries with small populations and small economies cannot provide sufficient education facilities, especially vocational training. Their labour forces are ill-equipped to cope with the modernization of the economy, so it does not go ahead. In almost every sphere of life many African states suffer from diseconomies of scale. Small political units make weak, dependent national economies which are easy prey to a neo-colonialism which promotes a world economic system in which Africa is near the bottom of the pile.

170

The lesson is well known within Africa and, having failed to take up Kwame Nkrumah's challenge to unite, many Africans have spent the first generation since independence setting up regional economic groupings in attempts to overcome the colonial inheritance of political balkanization and an inferior place in the world economic system. Many different regional groups have been formed to face the problems of diseconomies of scale and to promote intra-regional trade which is perhaps the most effective means of breaking down the cheap-raw-material-export/expensive-manufacturing-import cycle, of which most of Africa is a victim. There have been failures among these groupings but some have survived, though progress has been very slow.

At independence the most promising grouping was in East Africa. During the First World War British and Commonwealth forces occupied German East Africa, without completely defeating the small German army there, and at Versailles were rewarded by being given the territory under League of Nations mandate. The British then administered the three contiguous territories of Kenya, Tanganyika and Uganda as well as the offshore islands of Zanzibar. The opportunity was taken to create progressively a wide range of 'common services' for the four territories.

The idea of amalgamating the British protectorates in East Africa goes back at least to 1899, and closer co-operation possibly leading to a federation was encouraged by the British from the end of the First World War. From the beginning of the century common services were introduced by amalgamating the services of the individual colonies. By 1910 there was a Court of Appeal, common postal services and Currency Commissioners. A Customs Union between Kenya and Uganda was fully operational by 1917, and ten years later it was extended to Tanganyika. An autonomous Currency Board was set up in 1919. In the 1920s the range and depth of common services was expanded, and the services were expanded to include Tanganyika. For example, the common postal service became Post and Telegraphs and was extended to include Tanganyika in 1933. The East Africa Railways and Harbours and the East African Meteorological Service were created. A veterinary research organization was set up in 1939, a common Income Tax service in 1940. These common services grew up despite a lack of enthusiasm within East Africa for political federation in the 1920s, and without Federation the task of bringing together existing services was administratively difficult. Federation was a political issue disputed by the different interested parties inside and outside East Africa. African interests, the white settlers, mainly in Kenya, the Indian community and the Zanzibaris each took different perspectives, whilst the British Colonial Office, differing in enthusiasm as governments came and went, had the Permanent Mandates Commission of the League of Nations snapping at its heels, anxious to safeguard the rights of the people of Tanganyika. The Governors of the East African colonies met annually from 1930. The Second World War gave great stimulus to inter-territorial co-operation and at the beginning of 1948 Britain set up an East African High Commission to oversee the administration of the services shared by the British East African territories. Its powers covered the following common services: customs and excise, income tax, defence, and research and higher education, in addition to the self-financing railways and harbours, posts and telegraphs, and airways.

When independence came under serious consideration at the beginning of the 1960s, political federation was not a serious option, but the common services were put under close scrutiny as to the benefits they conferred on each individual territory. A Commission concluded that:

> the common market benefited East Africa as a whole, that its working could be validly criticized for producing an unequal distribution of benefits, rather than imposing losses on any member, and that no territory would be likely to gain from withdrawing ... The contributions the common market arrangements can make to economic growth are likely to be greater in the future than in the past.

> (East Africa Commission 1960)

With this endorsement it was decided to continue the common services and to replace the over-arching British High Commission organization in December 1961 (at the independence of Tanganyika) with the East African Common Services Organization under the East African Common Services Authority, consisting of the principal elected ministers of the three territories, in due course the three heads of state.

Julius Nyerere, leader of Tanganyika, the first of the territories to approach independence, offered to delay that independence until the other territories were ready. The offer reflected concern that the co-operation built up under the colonial adminstration would be put in jeopardy if the four territories were allowed simply to go their own ways. The offer was rejected, Tanganyika attained independence in December 1961, Uganda in October 1962 and Zanzibar and Kenya both in December 1963. As they had devised elegant but short-lived bespoke constitutions for the individual territories, the constitutional planners also made provision for an East African-wide administration to look after the existing common services and to provide a platform for further development.

In June 1963 the political leaders of the three countries (the independence of Kenya was assured but not yet realized) pledged themselves to 'the political Federation of East Africa'. They argued that the valued common services suffered from the lack of central control and that the time had come to create the central political authority. In 1967 the three countries came together formally in the Treaty of East African Co-operation to set up the East African Community (EAC). The EAC was designed to promote intra-community trade, and to administer and develop the wide range of common services that had been built up. There was also hope that the Treaty would prove the way forward towards realizing the aspirations expressed by the leaders of the three countries in 1963, and that in due course political federation would follow.

The 1967 Treaty was in effect the zenith of East African unity. Within ten years the East African Community had collapsed and the vision of political federation was yet another African mirage. Some of the causes of the collapse had long been present. They had been the matter of political debate and had been glossed over in the pre-independence report with assertion that, whatever grievances individual countries had, none would be better off if the central administration of common services was discontinued. Tanzania and Uganda were convinced that Kenya derived most benefit from the Community, and unfairly at their expense. The benefit was in the disproportionate share of trade, including the lucrative tourist trade, and the location of industry within East Africa. There were mechanisms for compensating Uganda and Tanzania under the Treaty but these were seen as inadequate because no amount of compensation could make up for losing individual commercial or industrial enterprises which generated growth within a national economy with spin-offs and a beneficial spiral of cumulative causation. Geographically Kenya was very much at an advantage at the centre of the Community. Its port, Mombasa, was the natural port for all Kenya and Uganda, and the northern part of Tanzania around Arusha and Moshi. Its capital, Nairobi, was the city nearest the geographical centre of the Community and its largest urban centre. Left to its own devices, incoming industry would choose to locate either at the chief port, Mombasa, or at the main local market and most central distribution point, Nairobi. With its large European population and modern infrastructure Nairobi had an added attraction for incoming commercial enterprises mindful of the preferences of its imported management. Any serious attempt on the part of the Community to direct the incoming enterprises away from Nairobi could well kill off the very iniative East Africa was so keen to attract. The political geography which had created the boundaries and largely determined the routes of the main trade arteries placed the Community at a disadvantage. Had there been no international boundaries within East Africa and no national sovereign governments, not only to cry foul but to threaten realistically to secede from the Community, things might have been different.

These matters arising from the basics of political geography were exacerbated by other factors. There was increasing ideological friction between the capitalism of Kenya and the socialism of Tanzania. These political differences were directly reflected in the locational choices made by incoming commercial enterprises. They were for the most part capitalist, often implants of multi-national corporations. Kenya's political ideology was the most in harmony with their own business culture and the firms saw in Kenya an economic environment with which they were familiar, but in Tanzania and in Uganda under Obote political environments of which they were suspicious. By the same token it was in Tanzania that investments from socialist countries were made, but they were far fewer in number.

A third major factor was the military *coup d'état* in Uganda in February 1971 which overthrew Obote and brought Amin to power. Nyerere of Tanzania, a close ally of Obote, deplored the *coup* and refused to deal with Amin personally. As the Community was formally headed by the three heads of state this made for difficulties. The spirit of co-operation dampened, the administration of the common services, the Community's central strength, became less efficient. A small outward sign in 1971 was that a visitor on internal East African Airways flights between Entebbe and Nairobi, and Nairobi and Dar es Salaam, could not buy in-flight drinks with any East African currency, only sterling or US dollars.

Things began to fall apart. The end did not come until 1977 when the EAC was disbanded as Tanzania closed its borders with Kenya. Two years later Tanzania was at war with Uganda in a successful attempt to overthrow the military regime of Amin. It was well into the 1980s before stability was again established between the three former East African partner states with their international borders open and the assets of the EAC finally shared between them. A Community based on long-standing common services and economic co-operation had foundered essentially because of an inability to co-operate politically. Differences in political ideologies were heightened by the different national sovereignties to make the problems of dealing with the inevitable economic bickering impossible. Had there been political will on the part of the three governments to make the Community work by insisting it overcome its economic problems it may well have gone on incrementally towards lasting co-operation.

French colonial groupings by other names tended to survive independence in part because independence came simultaneously to territories, many of which needed economic support, often given by France in its unique neo-colonial role. *L'union Douanière et Economique de L'Afrique Centrale* (UDEAC) succeeded AEF as a customs union in 1966 and also aimed to promote economic development of member states. It is less ambitious and its services less centralized than first conceived but it has survived, although with few positive achievements. *Organization Commune Africaine et Mauricienne* (OCAM), formed in 1965, has had many defections and now concentrates on technical and cultural co-operation. *Communauté Economique de L'Afrique de l'Ouest* (CEAO), the successor to AOF, was formed in 1974 and aims to lower customs duties between members and promote economic development, aims shared by the Economic Community of West African States (ECOWAS), to which all the CEAO states belong.

ECOWAS groups fifteen West African states from British and French colonial backgrounds plus the former private freed slave colony of Liberia. Its aims are to move trade towards customs union and to promote regional

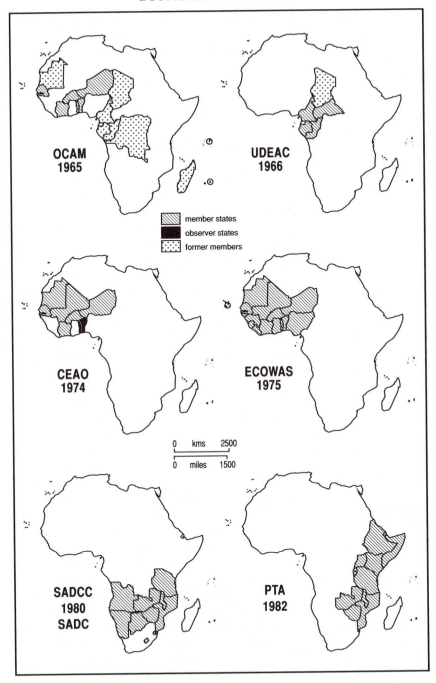

*Map 28* African economic groupings

co-operation through commissions for agriculture, industry, energy, trade, transport and telecommunications. Progress has been chequered and closely related to the prosperity of Nigeria, the dominant economic and political power within ECOWAS. The grouping was set up by the Treaty of Lagos in 1975 and the permanent secretariat is based in Lagos. It has four specialist commissions which deal with trade, industry, transport and communications, and social and cultural affairs. The economic size and strength of Nigeria relative to the other member states presents ECOWAS with a major problem, whilst the different colonial traditions are fault lines which are difficult to eradicate.

ECOWAS developed a political dimension in August 1990 when it created a multi-national military force (ECOMOG, the ECOWAS Monitoring Group) in its name to intervene in the desperate situation in Liberia. The main initiator, inevitably, was Nigeria whose contribution to the expedition was by far the largest. Other ECOWAS members contributed, even the tiny Gambia, whose President Jawara was strongly in favour of intervention, particularly during the period of his Chairmanship of ECOWAS, but some did not. Foremost among those not participating in the armed intervention was Ivory Coast, itself contiguous with Liberia. The lack of unity within ECOWAS in this matter has not helped in the success of the operation, which has not brought peace to Liberia, and to some extent has further muddied already murky waters. The expeditionary force occupies Monrovia, surrounded by the different warring factions. From there it launches punitive military sorties with air support, but the intervention seems to have lost its way. Whilst ECOWAS acted positively as a political organization, fulfilling a role perhaps more appropriate to the OAU, the intervention has emphasized the lack of unity within the organization even on this one issue. Success in the Liberian operation is a long time coming but the Nigerian military government has persevered. Recent moves by the various rebel forces, resulting in refugees flooding into Sierra Leone and the Ivory Coast, could have an outside chance of uniting the ECOWAS member states and success of the whole operation could follow. In that case ECOWAS would be transformed by sharing a common political purpose which could lead to further political successes. Nigeria's standing would be greatly enhanced. But that result of the most optimistic scenario could be damaging, because, added to its easy economic dominance, ironically, Nigeria's political importance might become so great as to cause other member states to shy away and so endanger he very survival of ECOWAS. The economic achievements of ECOWAS are modest but the extension into the political realm via Liberia is a fascinating development for the future direction of the organization. On the other hand the 1994 *coup* in the Gambia was partly an unforeseen consequence of military intervention in Liberia and by extension ECOMOG could hold a threat to the survival of the current military government in Nigeria.

The Southern Africa Customs Union (SACU) preceded the Union of South Africa, being set up in 1902 by the British colonial authorities. It included, as well as the four South African colonies, the three 'High Commission Territories' of Basutoland (Lesotho), Bechuanaland (Botswana) and Swaziland, and Rhodesia (Zimbabwe). The extension of SACU beyond the Union anticipated the inclusion of the other territories within the Union when conditions were ripe. In the event Rhodesia opted for 'responsible self-government' in 1923 and, despite intense pressure from South Africa, Britain could not bring itself to commit the High Commission Territories to South African rule. The white settlers in Rhodesia, having declined union with South Africa, chose to take an independent line to develop their economy behind protective tariff walls against South African competition. But SACU was extended to South West Africa (Namibia) when it became a League of Nations mandate administered by South Africa at the end of the First World War. The High Commission Territories eventually achieved independence from Britain in the 1960s and from the beginning of 1970 a renegotiated SACU of South Africa, Botswana, Lesotho, Swaziland and South West Africa came into being. It was a *verligte* (enlightened) agreement in South African terms, being generous to the smaller states by increasing the percentages of the overall customs revenue they received, making up for the former British grants-in-aid. In reality it was changing one colonial paymaster for another, an arrangement that very much suited the South African government. SACU is still in operation though the economic ties between the states have been loosened – for example, by Botswana establishing its own currency, the Pula, independent of the South African Rand in 1976.

Superimposed over the SACU states, and extending to include Angola, Mozambique, Zambia, Malawi and Tanzania, is the Southern Africa Development Community (SADC). As the Southern Africa Development Co-operation Conference (SADCC) it was set up in 1980 by the front-line states in the fight against apartheid, to encourage economic development and co-operation independent of a hostile South Africa. SADCC was a political creation to counter attempts by the apartheid regime in South Africa to extend its hegemony over the rest of southern Africa. Of these the most prominent was the concept of a Constellation of States put forward by the Botha government in 1979. This postulated co-operation between all the states of southern Africa in permanent orbit around a central, economically and politically dominant, but generous, South Africa. The implied static relationship did nothing for the aspirations of the SADCC states and was ideologically unacceptable to them, condemning them to permanent subservience to an abhorrent racist regime which offered not much more than crumbs from the table of the rich man of the sub-continental region. The independence of Zimbabwe in April 1980 was vital to the setting up of SADCC, though the preparatory work had begun earlier with the Arusha conference of 1979. Zimbabwe was the key piece in the geographical jigsaw of southern Africa, commanding communication routes between the other states of the region, particularly between the land-locked states and the ports of southern Mozambique.

During its first decade, transport was the main focus of SADCC's efforts. SADCC brought together the southern African countries to work out priorities for transport development and rehabilitation within the area, in effect to draw up a prioritized shopping-list to take to potential aid donors. In this way the organization served well as an aid channel, being particularly successful in attracting aid from the Nordic countries to co-ordinated transport projects. Much of the work of the SADCC in the 1980s was negated by the South African destabilization of its neighbours. Roads, railways, bridges and harbour installations were sabotaged by the white South Africans and their proxies, the MNR in Mozambique and UNITA in Angola as the Botha regime strove to enforce political and hegemony over the region. Angola is still suffering the effects of an on-going civil war where the main anti-government protagonist is UNITA, for so long nurtured by white South Africa and the United States. The Angolan civil war keeps the Benguela railway closed and so reduces the transport options available to Zambia and other SADC states. Even if the war were to end shortly, the railway cannot be reopened without a massive programme of rehabilitation. Although SADCC developed other commissions, transport remained its dominant concern throughout the 1980s. This policy was understandable in as much as a basic infrastructure is essential to any future co-operation, but the lack of progress in other directions was discouraging.

The changes in South Africa from February 1990 led SADC leaders to state in 1992 that they would welcome participation in SADC by a majority-ruled South Africa. This represents a fundamental change in the aims of SADC. Until now one of the main thrusts of SADC has been economic development *independent* of South Africa: indeed that was the political *raison d'être* for the creation of SADCC in the first place. Whilst it is natural to embrace majority-ruled South Africa, it could be an embrace deadly to the independent aspirations of the other member states. South Africa dominates the region economically, with 20 per cent of the land area, 40 per cent of the population, and 80 per cent of the wealth (GNPs). With the abolition of international sanctions against South Africa in 1993, followed by majority rule in 1994, the South African economy, so long in deep travail, shows signs of resurgence. South Africa could become even more economically dominant within the region and a regional 'brain-drain' is already evident. The economic strength of a friendly South Africa might be more difficult for its neighbours to withstand than the brutal hostility of the apartheid state. Johannesburg and the Rand are the economic hub of the region in terms of commerce, industry and transport despite the apartheid years when SADCC fought to assert independence from South Africa. The ports of South Africa could well continue to attract traffic from the SADC states despite the rehabilitation of the Mozambique ports and their feeder transport routes. South African ports have spare capacity, are well equipped and are poised for further modernization and expansion. On the other hand SADC could benefit from development planned with the South African economy as the driving force. Removal of the threat of political hegemony makes the prospect of economic domination more palatable. Much remains to be worked out for the future of SADCC and the participation of South Africa represents a major change of direction.

Parallel with SADC is the Preferential Trade Area (PTA) formed in 1982. It embraces a wider area than SADC, extending from southern to eastern Africa, from Lesotho to Djibouti. It aims to promote intra-regional trade and joint action by member states for production of certain goods, services, including financial, and resource development in an attempt to integrate the national economies into a regional economic community. Intra-regional trade as a proportion of the total trade of the member states is very small, because exports from member states are mainly primary products and raw materials whilst imports are mainly manufactured goods. The main trading partners of the member states are in the industrialized first world. This is the fundamental problem African countries face, especially as the terms of trade have moved consistently against primary products and in favour of manufactured goods during the whole of the post-independence period in Africa. In a difficult first decade the PTA has made little progress but some structures put in place could lead to useful developments, particularly when membership is extended to South Africa. The price paid could be dominance by one power at the expense of the others, but at least it would be an African power.

Independent economic development in Africa remains a dream. Political balkanization of Africa made economic dependence and neo-colonialism inevitably part of the African inheritance. In the world economic order Africa has been given a menial and subservient role. The main trading partner of almost all African countries is the former colonial power. The trade itself for most African states comprises exports of raw materials, minerals and cash crops, and imports of manufactured goods. Trading with industrialized countries in itself puts African countries at a disadvantage. Exporting raw materials and importing manufactures compounds that disadvantage. The direction of trade and the content of trade is wrong for Africa. On the other hand the prospect of African states trading mainly with each other is distant because of the similarity of the exports they generate and the imports they demand. African economies need to be restructured so that they are better placed to trade with each other.

To set Africa on the right economic path the creation of economic communities of states is probably an essential first step, but it is one fraught with difficulties. As a result the post-independence communities which have been created have either failed, like the EAC, or, like ECOWAS, SADC and the PTA, have achieved very little. Economic difficulties in the way of co-operation, such as natural imbalances between states, are exacerbated by political differences which, because they are often between sovereign states, are extremely difficult to overcome. With the failure of the EAC as a pertinent warning it is not surprising that the PTA, which involves the same East African states plus many others, has not made much progress. One of the causes of failure of the EAC was the imbalances between Kenya and the other two member states and the inability to create adequate means of compensation. Within the PTA for its first decade a single polarity was replaced by a dual polarity, Kenya being matched by Zimbabwe in terms of economic development. But matched is not the same as balanced and the larger PTA contains an even wider range of national economies with much greater differences between levels of economic development and modernization. When South Africa is included within the PTA the differences are even greater. Little wonder the PTA has yet to progress towards becoming a free trade area for it is then that its problems will really begin. Yet a free trade area is a necessary step along the way to fuller economic integration.

Africa badly needs to develop intra-African trade as a means of getting off a world trading treadmill that condemns the continent to long-term poverty. To meet African demands manufactured goods must be produced in African countries. To do that African raw material resources must be used in Africa where the benefits of manufacture will accrue to Africans. There is also an urgent need to create the right intra-continental infrastructure to support such changes. These objectives must be met despite the political balkanization of Africa and indeed as a means of overcoming that excessive division. Political balkanization, economic fragmentation, the direction of African trade and the content of African trade are all part of the African inheritance and present formidable obstacles on the road towards political and economic independence and prosperity.

# 16

# INFRASTRUCTURAL DEVELOPMENT

Infrastructural development is of crucial importance to the creation of modern economies in Africa and in particular to the evolution of intra-African trade which is the best means of freeing Africa from its neo-colonial bondage. Parts of Africa inherited from colonial times a well-developed infrastructure, others did not. But even where there were good railways, roads, airports, sea ports, telecommunications, public utilities and services they were specifically created to meet the needs of colonialism. At independence there was enormous scope for improving African infrastructure absolutely and for reorientating existing infrastructure to serve better the needs of independent as opposed to colonial development.

Transport development in Africa has three distinct, if overlapping, phases: pre-colonial, colonial and post-colonial. However, most models of transport development in Africa, such as the 'classic' Taaffe, Morrill and Gould model of 1963, are colonial in concept and scope and so forfeit historical accuracy and current relevance. They also pay little attention to the fact that routes, especially those involving heavy capital investment, are developed with a specific goal in mind. Roads and railways are not built aimlessly across isotropic surfaces through empty interiors. Such models lack a logical dynamic and display weaknesses, enhanced by uncritical reception and dissemination, which lead to fundamental misconceptions about African transport development. There is in particular a need to build a model of transport development in post-colonial Africa to take into account the different considerations which influence decision-making in the independent situation. The pre-colonial phase also needs to be sketched in and the colonial phase rewritten to accommodate the criticisms outlined above.

Before colonial times there were rich agricultural areas in Africa, concentrations of population, some towns and cities and worked mineral deposits. People moved between such places along well-defined routes, carrying trade goods over considerable distances. On the southern edge of the Sahara, Timbuctoo and Gao stood at crossroads of trade. Across the Sahara came Mediterranean products and from the south came gold, ivory and slaves. Trans–Saharan and east–west savannah routes long predated European exploration. To reach Timbuctoo Gordon Laing followed well-established caravan routes across the desert from Tripoli. René Caille obtained passage on a boat, one of many sailing regularly up and down the navigable inland Niger, to Timbuctoo. Caille returned to Europe via the very old caravan route to Morocco. The route from Bagamoyo to Lake Tanganyika was not hacked out by Burton and Speke, but was an old-established Arab trade route, in a sense colonial as it had been developed and was primarily used to exploit ivory and slaves. Kampala–Mengo as a node long pre-dated the arrival of Speke and Grant. Zimbabwe traded its gold with the Arabs of Sofala for centuries before the Portuguese sailed around the Cape. Patently, pre-colonial Africa was not an empty continent, but that was what the armchair geographers of Europe thought it was, and the map-makers drew 'elephants for want of towns' (Dean Jonathan Swift). The transport geography model-makers of the 1960s, with far less excuse, made the same mistake.

When the Europeans came they established trading posts on the coast and, by offering higher prices for gold, ivory and slaves, contributed to the decline of savannah trade centres such as Timbuctoo. Early European traders did not venture far into the interior. African middlemen developed effective routes to the coast. When the Europeans did venture inland it was usually to pre-existing African towns, often for military purposes. For example, in present-day Ghana in 1874 the British went inland to attack Kumasi, capital of the troublesome Ashanti. They built a military road from Accra to Kumasi as they went, renewed it in their campaign of against Prempeh in 1895, and in 1923 built a railway along the same route. In Nigeria British penetration was also to pre-determined points, to pre-existing African cities such as Ibadan and Benin and later to Kano and Sokoto. Colonial transport routes were built from colonial ports, often to pre-colonial places, following well-trodden paths and trading routes and were not developed in a virgin wilderness.

In southern Africa the colonial phase may be usefully sub-divided into pre-industrial and industrial, the latter dating from 1870, when diamonds were first discovered at Kimberley. Pre-industrial colonial transport was mainly ox-wagon, a leisurely but effective means of transport from the ports to the few inland centres. Accounts of such travel survive in sources as well known as Livingstone's *Missionary Travels* (1857). The interior had little to attract investment in any means of transport other than ox-wagons, which were adequate for carrying the wool, hides and ivory produced in the interior. Penetration by railway was discouraged by the outward-facing Great Escarpment and, in the south-west, the ranges of the Cape Fold Mountains, also parallel to the coast. By 1870 there were just 69 miles (110km) of railways in South Africa, mainly linking Cape Town with Stellenbosch, Paarl and Wellington across the Cape Flats, and one mile at the port of Durban.

Transport development in southern Africa was transformed by the discovery of diamonds at Kimberley in 1870. The mines were rich, localized and at the heart of the sub-continent. Suddenly there was a need for modern transport *and* a means of financing it. Existing railways were 'nationalized' and under regional political pressures railways were started towards Kimberley simultaneously from Cape Town, Port Elizabeth and East London. A new 'Cape gauge' of 3ft 6ins (1.065m) was adopted to ease the engineering and financial problems of breaching the mountains. But it was not until November 1885 that the railway from Cape Town and Port Elizabeth reached Kimberley, having met earlier at De Aar junction. Investment in railways came only after a worthwhile goal had been identified and the railways were aimed at that one specific place. The fact that Kimberley happened to be at the geographical heart of South Africa was the critical fact in the opening up of the country to modern transport.

The Witwatersrand gold-fields, discovered in 1886, provided more of the same. They were also localized and deeper into the interior. The first railway line to reach Johannesburg was an extension of the Cape line, via Bloemfontein. Construction over the relatively flat high veld was quicker and cheaper than climbing the escarpment afresh. It was allowed to enter the Transvaal in 1893 only after considerable political and financial bargaining between Kruger and Rhodes. Although a railway had been built from Durban towards the diamond-fields it had only reached Ladysmith beneath the Drakensberg escarpment by 1885. Later this Natal line, extended towards the gold-fields, was held at the Transvaal border for political reason, to allow completion of the Transvaal's own non-British line from Delagoa Bay. On this evidence the reality of colonial railway construction was a heady mix of investment capital and interest rates, gradients and gorges, earthworks and construction costs, projected traffic, tariff rates, profits and politics.

From 1890 Rhodesia (Zimbabwe) became the next node in the continental interior. Although the short, direct Beira route was the looked-for means of access, the Matabele campaign of 1896 and the rinderpest pandemic, which disrupted ox-wagon transport at a critical juncture, led to the urgent mile-a-day extension of the spinal railway from Mafeking (Mafikeng) to Bulawayo, completed in November 1897. The Beira line slowly struggled through fevered swamp and over rugged escarpment to Salisbury (Harare) in 1900.

Further north the newly discovered mineral wealth of Broken Hill (Kabwe) and Katanga (Shaba) became the next nodes of attraction. They were reached by the spinal railway in 1905 and 1910 respectively. The all-Belgian line from Port Francqui (Ilebo) on the Congo river system to Elisabethville (Lubumbashi) was completed in 1926 in advance of the Benguela railway from Lobito, also to Katanga, in 1928.

The South African campaign against the Germans in South West Africa (Namibia) during the First World War led to the Cape network being connected from De Aar to the German rail system. This was eventually upgraded to Cape gauge throughout and integrated into the South African system. Colonial route developments in Africa, road and rail, were frequently in support of military action, whether in this case and in East Africa, against another colonial power, or in the case of Kumasi and Bulawayo, against Africans.

Southern Africa has by far the largest single rail network in Africa, over 20,000 route miles (32,000km) in twelve different countries, at the standard Cape gauge. The densest part of the network is in the richest country, South Africa, which has over 13,500 route miles (21,560km). Primarily built to serve the important mineral-based nodes the rail network also encouraged development between nodes. Hence the concentration of economic development along the Durban–Johannesburg corridor in South Africa, the Harare–Bulawayo axis in Zimbabwe and the Zambian 'line of rail'. The southern African rail network was a colonial creation initially designed to facilitate mineral exploitation and export and to strengthen white (specifically British) political domination. It was the means whereby European capital penetrated the region. It also became the infrastructural framework of the sub-continent, strongly influencing the distribution of economic development. In another role, in the 1980s it provided white South Africa with the means of exerting economic and political hegemony over neighbouring states despite important post-independence railways built specially to help break that dominance.

In other parts of Africa colonial development did not produce a rail network but a series of railway lines built from port to a specific inland goal, often a mine, sometimes a perceived strategic location important in securing colonial rule over Africans and against the threat posed by rival European imperialists. In East Africa three separate colonial lines from the ports of Mombasa, Tanga and Dar es Salaam were eventually joined to form a single network at a metre gauge. The Uganda Railway was built by the British for strategic colonial reasons to Port Florence (Kisumu) on Lake Victoria and later into Uganda itself. The Germans built two lines in their East African colony: from Tanga to the rich agricultural area near Mount Kilimanjaro and from Dar es Salaam to Kigoma on Lake Tanganyika, with a branch which was later extended to Mwanza on Lake Victoria. The Mombasa and Tanga lines were linked for military purposes in 1916 when South African forces under General Smuts, fresh from linking the Cape and South West African systems, attempted to chase the German General Von Lettow Vorbeck out of East Africa. Significantly, the final link with the Dar es Salaam line deliberately to make a single network came after independence in 1964.

Elsewhere in Africa, the railways are essentially colonial in construction and purpose, except in Maghreb where a lateral line runs from Marakesh to Tunis. They are mainly short lines from the coast inland to mines, to larger towns or to rich agricultural areas. Their purpose was to facilitate exploitation of minerals and cash crops, and to provide ease of communication for the military authorities to control any anti-colonial situation which might arise in the interior. These lines have no lateral linkages, except in Nigeria where the lines from Lagos and Port Harcourt form a simple network entirely within Nigeria. Three other colonial railway lines, from Dakar, Abidjan and Djibouti do cross international boundaries, linking Senegal with Mali, Ivory Coast with Burkina Faso and Djibouti with Ethiopia. In Liberia, Guinea and Sierra Leone individual lines within each country were of different gauge, making difficult any possible lateral linkage.

Post-colonial or independent transport development in Africa began in 1894 when the South African Republic's (Transvaal) direct link with Lourenco Marques was completed in an attempt by the Boer republic to escape the British economic stranglehold. It was 'post-colonial' in that the Boers had in 1881 re-established their independence from the British and 'independent' because the decision-making for construction and finance of the line was done in Pretoria, which was not part of a colonial empire. Much the same happened in Rhodesia when, by decisions taken by a settler federal government in Salisbury (Harare) eager to maintain its economic independence from South Africa, a direct rail route to Lourenço Marques via Malvernia (Chicualacuala), avoiding South African territory, was opened in 1955. When Mozambique was about to become independent, white Rhodesia's direct links with the sea were threatened. The illegal Smith regime based in Salisbury decided to complete a direct rail link with South Africa via Beit Bridge in 1974, although it and earlier governments had delayed doing so for almost fifty years. As in the case of the Transvaal the Rhodesian initiatives were independent of the colonial power and were taken with a continental perception rather than that of imperial metropole.

The most spectacular post-colonial railway in Africa is the 1,050-mile (1,680km) Cape-gauge Tanzania–Zambia (TAZARA) railway from the Zambian Copperbelt to Dar es Salaam. It was planned specifically to help Zambia escape the clutches of white Rhodesia and South Africa. The decision to build was taken in post-colonial Lusaka for blatantly political reasons which did not apply in the colonial period when the traditional access route was under British rule. Given the hostility of the white south Zambia had no option but to seek alternative access to the sea. The bottom line was not railway profitability but political survival. Less ambitious and dramatic but also to give a land-locked state alternative access to the sea is the rail link between Malawi and the northern Mozambique port of Nacala. These lines reflect the fact that the decisions to build them were taken in independent continental capital cities of African (and Afrikaaner and white settler) land-locked states. They are not lines likely to have been built by colonial powers looking at Africa from the outside or even from colonial capitals such as Cape Town.

Other post-colonial railways have penetrated remote areas to assist regional development, for example, to Maiduguri in north-eastern Nigeria, to Packwach in Uganda and the Trans-Gabon railway. Some railways built in the post-colonial period, such as the Kasese extension in Uganda to serve the Kilembe copper mine, are colonial in concept. But most post-colonial railways are different in conception from colonial constructs and must be dealt with separately in any treatment of transport development in Africa.

Africa inherited from the colonial period few paved trunk highways, partly because of the inevitable time-lag in technological diffusion but mainly because of a lack of commitment to investment by colonial governments starved of resources by the metropolitan countries. An exception was the partially paved 540-mile (870km) *Strada Imperiale* from Assab to Addis Ababa which the Italians built to secure Ethiopia. Road-building gained momentum in the post-colonial period, but many trunk roads merely duplicate routes already served by rail, often from ports into the interior. Others connect places which were previously most accessible by river transport, for example, from Banjul along the south bank of the river Gambia to Basse Santa Su, the first 125 miles (200km) of which to Mansa Konko is tarred using a matrix of sea-shells rather than stone chippings. These roads represent an inevitable but extremely costly modernization of the trunk transport network without actually extending it beyond the colonial concept.

Another part of the African inheritance is that the land-locked states were left to build their own roads of access to the sea with the co-operation of one or other of their seaboard neighbours. Because those states are no longer parts of imperial empires, alternative routes have often necessary in the post-independence era. So Mali had to construct a tarred road to the Ivory Coast railway as an alternative to the Dakar railway. In the late 1960s the road from Kampala to Mombasa was finally tarred to give a brief period of smooth journeys to the coast before the pot-holes really developed. In Ethiopia a 150-mile (250km) branch of the *Strada Imperiale* was, with foresight, extended to the port of Djibouti.

Increasingly, too, tarred roads are used to reach those parts previously not connected by modern transport to the capital city, for example, the Trans-Gambia Highway from Dakar across the Gambia to the previously remote Casamance region of Senegal, or the road west from Lusaka via Mongu to Barotseland, or the road from Nairobi north to Lake Turkana. Such roads have political as well as economic motivation and help bind to the centre remote parts of the states of Africa.

Nevertheless the pattern of international transport in Africa is still basically colonial. Its main components connect port with interior. Even where the initiative comes from the interior, for example, from a land-locked state, the interior-to-port connection is exactly the same. It reflects the continuing pattern of African trade, which is mainly with industrialized countries of Europe and America rather than intra-African. Most African

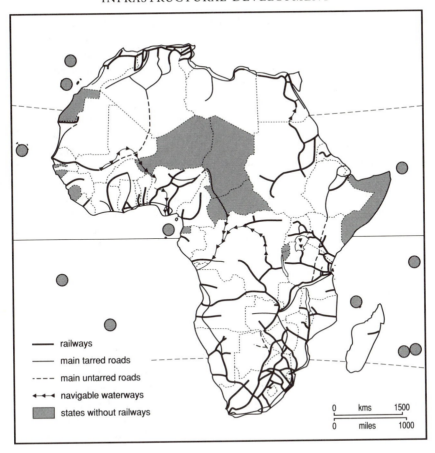

*Map 29* International transport in Africa

countries have not addressed the chicken-and-egg problem: direction of trade influences transport routes which influence direction of trade. They could deliberately build roads to connect with their neighbours but have not done so except to gain access to the sea. The painfully slow progress of the long-mooted trans-African highway project illustrates the lack of urgency felt in Africa for such links. The Trans-African Highway Authority was formally inaugurated in 1981, ten years after it was first proposed by the Economic Commission for Africa (ECA). Thirteen more years on, the project is nowhere near completion and still lacks effective enthusiastic champions, national governments willing to find the investment necessary to get the job done.

If Africa is ever to free itself from a world trading system in which it occupies a subservient and dependent role it must begin to develop an international continental-wide infrastructure capable of materially assisting the development of alternative trade patterns. Intra-African trade can only be expanded when the means of pursuing that trade are in place. Once in place the roads and railways must be fully maintained. Road surfaces and permanent ways must be repaired, lorries, locomotives and rolling stock regularly serviced and running operations performed efficiently.

Most African countries have developed airlines since independence. At independence for many countries, world aviation was only just taking off and the colonial inheritance was very meagre. Airports were built to accomodate successive generations of inter-continental jet aircraft, and fleets of incredibly expensive aircraft were acquired. Many countries bought the biggest and most expensive to ply the long-haul routes to Europe and North America. The emphasis was usually placed on these inter-continental enterprises rather on intra-African connections. Where there were few scheduled flights between East and West, North and South Africa, they were usually flown by European and American carriers, such as the former once-weekly, always over-booked, Sunday afternoon Pan-Am flight from Nairobi and Entebbe to Lagos, Robertsfield (Liberia), Dakar (and New York). The likes of Air Uganda and Air Kenya put their main resources into regular flights to and from Europe. Often, to fly from East to West Africa, it is still quicker, more convenient and cheaper to fly via London or Paris. At the height of Amin's reign of terror Air Uganda flew weekly from Gatwick and Stanstead to Entebbe loaded with goodies to keep the brutish military happy. More benign, the tropical fruits and vegetables flown to European markets 'by air for freshness' perpetuate the trade in African cash crops, which are traded at such disadvantage for the manufactures of Europe.

Infrastructure is much more than railways, roads and airways. Colonial governments built great hydro-electric dams in Africa, sometimes as parting gifts before independence, such as the Owen Falls Dam in Uganda where the Nile leaves Lake Victoria. Others, such as the Kariba Dam, had a different political motivation, built as a symbol of the unity of a white minority-ruled Federation of Rhodesia and Nyasaland. But many dams were not built in the colonial era and Africa's enormous potential for hydro-electric power was only fractionally realized. Although large dams are favoured by aid donors and recipient African governments alike, not enough have been built in the post-independence period to provide Africa with an essential prerequisite for economic development and modernization; and it is not only the large dams of the prestige project but the smaller dams, which may or may not produce electricity but which control rivers, tame wild riverine regimes, conserve soils and provide irrigation water. Contrast the intensity of dam distribution on South African rivers and tributary streams with the paucity of such constructions in so many other parts of Africa. The colonial inheritance was often the monster dam. There were too few of them but almost no smaller dams, less costly in construction and displacement costs and collectively every bit as productive as one great show-piece.

Other infrastructural inheritances from the colonial period have also been inappropriate, even if originally constructed with the best of motives. A large, expensively equipped and staffed University teaching hospital, such as Mulago in Kampala was, arguably, not the most effective way to invest scarce resources in medicine and health care in Uganda. There is no denying the often distinguished contribution of such hospitals to medical science but are they the most effective way of extending health care to the greatest number of people in poor countries where even rudimentary facilities and expertise are not available over wide areas? Are the highly qualified doctors who emerge at the end of a long and costly Western-style medical training better able to minister to the needs of most rural dwellers than medical auxiliaries trained in a far shorter period and at a fraction of the cost? The government of Tanzania did not think so and embarked on a radical reform of its medical training programme to give emphasis to getting relatively cheaply trained auxiliaries into the field and to drawing up a structured clinic and hospital service. Mulago was another British parting gift but was it geared to the real medical and health care needs of Uganda? In an ideal world Africa should have its large university teaching hospitals but should they have priority over many rural medical centres and clinics which would bring basic low-cost medicine and health care within the reach of many more people? Symbolically the large hi-tech teaching hospital is like the vast hydro-electric dam, extremely valuable but not perhaps the most effective way of meeting Africa's most pressing and immediate needs. The African inheritance is here not only the bricks, mortar and concrete of the great symbolic structures but more an attitude, a well-intentioned but alien approach which might not be the most appropriate for Africa. This is not to say that second-best will do for Africa. It is more a matter of taking a hard look at the problems to be faced and the means of tackling them. Resources available to African countries are severely limited and need to be used in the most efficient and efficacious manner.

Another aspect of infrastructure is education and much the same criticisms can be levelled at the African inheritance. At the level of the basics of education, literacy, too many African countries faced a situation at independence that was dire. The position has improved steadily since then but in many African states, notably those of the Muslim Sahel, literacy levels are still low and appallingly so for women. The provision of higher education in the colonial era was very sparse and therefore extremely patchy. At one time the perceived needs of British-ruled Africans in the whole of southern and East Africa were serviced at Fort Hare in South Africa. Later Makerere University College was developed in Uganda whilst Ibadan University served British West Africa. These institutions were British in tradition and British in curriculum. Makerere proudly displayed photographs of its distinguished

189

former alumnus Milton Apolo Obote playing the part of Julius Caesar in the Shakepearean tragedy. Perhaps there were political lessons contained in the play that Obote took with him into Ugandan politics, but the question of relevance has to be raised. This tradition was passed on, in part to fuel the debate over the 'diploma disease', the acquisition by many Africans of qualifications irrelevant to the pressing needs of their countries.

Much of the present African infrastructure was part of the colonial inheritance. As such it was orientated towards serving colonial ends rather than independent African interests. More seriously, since independence infrastructural development in Africa all too often has followed lines laid down in the colonial period and is not always appropriate to the best interests of the independent state. The colonial way was mainly to treat each individual colony as a self-contained unit, except where two colonies of the same colonial empire were contiguous. Infrastructural development in modern Africa needs to incorporate much more international co-ordination. Meaningful independence in Africa can only be fully expressed where African states co-operate with each other for their common good. That can only be done when African infrastructure is developed to facilitate fully that co-operation.

# 17

# CONCLUSIONS

Africa is the most disadvantaged continent in terms of poverty, political unrest, quality of life and human suffering. Many, though by no means all, of the ills of Africa derive directly or indirectly from its colonial inheritance. The effects of the recent colonial past are not easily put aside. In the generation since independence conditions in Africa have steadily got worse, as exploitation has continued, and in many spheres are now at an all-time low. Can anything be done to exorcise the effect of past evils, to halt the decline which seems doomed to end in human catastrophe of unprecedented scale?

The clock cannot be put back in respect of political geography, which is a root cause of so many African problems, but the most serious effects can be ameliorated. Any worthwhile schemes for achieving this will be costly and at present are well beyond the means of most African countries. To put African countries in a position to help themselves, the chronic debt burden of Africa must be lifted. This would wipe the slate clean and enable a fresh start. Beyond that, exploitation of Africa must end and Africa has to be encouraged and materially assisted with carefully targeted aid, not in the conventional sense of 'hard' or even 'soft' loans, but funds made available more in the way of putting right past wrongs, some form of reparation from the industrialized world, particularly, though not exclusively, from those countries which formerly held African colonial empires. Such payment on the part of the industrialized world would be more a matter of enlightened self-interest than anything else.

Political balkanization may be in part ameliorated by co-operation on a continental or sub-continental scale in Africa. Existing organizations for economic co-operation such as ECOWAS, the PTA and SADC should be encouraged and assisted. They are existing African initiatives and as such are likely to succeed better than any initiative made from the industrialized countries through whatever international agency, including the World Bank and the International Monetary Fund. Lessons learned elsewhere in the world about overcoming the difficulties experienced in achieving international economic co-operation should be passed on through assisted

191

transfer of expertise by means of specialist educational training and, where appropriate, ex-patriot secondment.

A major long-term contribution would be in the form of infrastructural development in Africa, particularly with a view to promoting intra-African trade and economic complimentarities. Such developments call for international co-operation and co-ordination in projects which will be costly and will yield no short-term tangible results. Road-building programmes could ensure better communication between African countries, but they would have to be accompanied by measures to secure road (and rail) maintenence as well as the adequate servicing of vehicles, locomotives and rolling-stock. There is still a tendency in Africa for international boundary lines to repel modern economic development and there are, for example, few cross-border paved roads compared with the intensity of networks developed within national boundaries.

Until more international links are in place the development of intra-African trade will be hampered. For example, there are few modern surface communication links (one-metre gauge railway and unsurfaced roads only) between the extreme west coast states of Mauritania, Senegal, the Gambia, Guinea Bissau, Guinea and Sierra Leone, and states to the east, Mali, Ivory Coast and Liberia. There are also no paved roads or railways between Nigeria and Niger and the states to the east, Cameroon and Chad. These two 'fault lines' between communications blocks happen to coincide with long-standing political divides which have frequently experienced war, as currently on the Sierra Leone/Liberia border and in the constant minor conflict on the Nigeria/Cameroon border. Does the lack of modern communications across these zones reflect the political tension or does it cause tension? Chicken or egg? Liberia, so long in the grip of ferocious civil war, is almost completely isolated from Sierra Leone and Guinea to the west and from Ivory Coast to the east. Chad has only one paved road connection with the outside world (though not to the sea), whilst the Central African Republic depends on the Zaire River for its external surface transport contacts. The isolation of these states is all the more serious because they are land-locked. Farther south along the west coast of Africa, Equatorial Guinea, Gabon and Congo each have no paved road or rail connection with any other state.

East Africa is isolated from central African countries in terms of modern surface transport. There is no paved road or railway west or north from Uganda to Zaire and the Sudan, and none from Kenya to its northern neighbours of Somalia, Ethiopia and the Sudan. The lack of international co-operation is well illustrated by the latter divide, as a paved road stretches for over 375 miles (600km) north of Eldoret in Kenya to within 20 miles (30km) of the Sudan border, where it meets an unsurfaced road for the 250-mile (400km) route to Juba. A paved road extends south for about 500 miles (800km) from Addis Ababa to Moyale on the Kenyan border where it meets a murram road for the first about 300 miles (500km) on the route to Nairobi. Again these 'fault lines' in international communications in Africa coincide with civil war zones and wars which occasionally spill over the international boundaries, across which there are no easy modern communications. War-stricken Sudan, like Liberia, has no paved road or rail links with another state, the only scheduled service being by ship along Lake Nasser between Wadi Halfa and Aswan in Egypt. Somalia, another war torn country, is similarly completely isolated in terms of modern surface transport from all three of its contiguous neighbours, Djibouti, Ethiopia and Kenya.

Southern Africa as a whole is much better interconnected than the rest of the continent but there are some 'fault lines' limiting the modern communications network. One is between Tanzania and Mozambique where the 475-mile (760km) boundary, made up mainly by the River Rovuma, is not crossed by a single road of any description. On the other side of the continent Angola, another country afflicted by long-standing civil war, is also largely isolated. The Benguela railway which connects with Zaire has been closed since 1975 and only in the south, where a paved road connects with Namibia, is the isolation from neighbouring states broken.

The many 'fault lines' in intra-African communications have many causes. In addition to those mentioned above are those which are largely due to the great deserts. The Sahara is not fully crossed by any paved road or railway. Along the west coast the Moroccans have constructed a paved road for the full length of Western Sahara, 855 miles (1,370km), connecting with the Mauritanian port of Nouadhibou, which is isolated from the rest of Mauritania. In Algeria paved roads penetrate deep into the Sahara but still leave large gaps crossed only by sandy tracks on which motor transport is 'subject to special regulations'. In the south the Kalahari Desert isolates Namibia from Botswana and the rest of south-central Africa. The communications 'fault lines' already noted in equatorial Africa are also largely due to physical conditions, namely tracts of tropical forest.

The impediments of physical geography can be overcome but at a cost, which in many cases will be very high. More difficult is to overcome the blocks on international surface transport and communications caused by political factors, particularly cross-border hostilities and civil war. Even where paved roads and railways are in place they are useless when closed by war. So fundamental to establishing good intra-African transport and communications, which themselves are a necessary prerequisite for developing intra-African trade and economic co-operation, is the need to establish political stability and end all international and civil war on the continent, quite a tall order.

The problems of Africa can only be solved in the long term by Africans. Outsiders have for far too long interfered in African affairs and must take a great deal of the responsibility for the parlous state they are now in. Nevertheless the French, for example, are still inclined to meddle in Africa, as the 1994 experience in Rwanda has shown. They claimed their intervention was solely for humanitarian reasons at a time when there was a real need for such intervention because the United Nations (UN) operation was unable to get off the ground. There was widespread disbelief in Africa and beyond that the French did not have some other agenda related to their earlier involvement with the about-to-be-deposed Hutu extremist government and the discredited Mobutu government in Zaire. In creating a safe zone in the south-west of the country the French were seen as protecting a government, army and militias who had advocated and executed genocide. The motive was perceived as the ever-present French need to play an active role in Africa, the only part of the world stage where they still, after years of persistant involvement, carry some weight. It is a self-important neo-colonial role to which the French cling with fading memories of an imperial past. The actual role they played was, in fact, much closer to their claimed humanitarian role than anyone anticipated, probably including the French themselves. Whatever the merits of the French intervention, in an ideal world it should not have taken place. There was a very real need for intervention on humanitarian grounds to try to prevent the genocide and supervise and police the relief effort spearheaded by the Non-Governmental Organizations (NGOs). But that intervention should have been by African states under the aegis of the UN or the Organization of African Unity (OAU). That did not happen, largely for want of money to fund the operation, even through the UN, but also because no proper mechanism existed to organize such an intervention on the part of the OAU.

It is essential in future that a mechanism be agreed by African states, under the umbrella of the OAU, so that there can be quick and effective African response to help avert human disasters such as the Rwanda crisis. Funding might appropriately be provided through the UN, which would allow the industrialized countries to contribute. But the mechanisms and the funding should be put in place, the latter as a UN contingency fund which can be topped up as needed, before the next crisis develops so that intervention can be both swift and effective. The problem that also has to be tackled in advance is the matter that has deterred OAU and UN action in the past, and that is the right to intervene in what might be described as a civil war. The UN presence in Rwanda had been established, at the invitation of Rwandan government, before crisis became genocide and the authority of the government was really challenged. The tragedy was that the opportunity to build from this base was not taken and that the UN presence was actually drastically diminished as the crisis deepened. African states must be prepared to take steps initially to prevent a crisis developing and, if that fails, to take the initiative to intervene. Industrialized countries, particularly former imperial powers, should resist the temptation to intervene directly but should be prepared to help defray the costs of an authorized African intervention through the UN.

Beyond the problems posed by Africa's many wars and the need to improve the continent's infrastructure in order to promote intra-African trade and co-operation there lies the fundamental problem of restructuring African economies so that they become interdependent and supportive of each other. Too many African economies are so similar to one other that they have little to offer each other for trade. Africa needs to develop national economies that complement each other and produce goods that other African economies would want to import. Most African economies are based on raw material exports which are traded for imports of manufactures produced in the industrialized countries. There is virtually no demand in Africa for copper ingots, which comprise about 98 per cent of Zambian exports, nor for the cobalt of Zaire, the chrome of South Africa, the bauxite of Guinea and the asbestos of Zimbabwe. Many African countries export unprocessed cash crops, such as cocoa, coffee, tea, groundnuts, palm oil, bananas, peppers and pineapples, but very few import such items.

The structuring of African economies that has taken place under pressure from the IMF has all too often placed renewed emphasis on such raw material production. Not only has the concentration on cash crops for export hurt Africa's ability to feed itself but it has also served to confirm Africa's place at the bottom of the pile in terms of the world economy. What is needed is a radical approach to Africa's economies so that they become supportive of one another and attempt to secure independent development. African economies might be weaned off cash crop production, concentrate agriculture on food crops for home consumption, and develop some manufacturing industry so that they have the wherewithal to trade with each other. Most individual African countries, for reasons of scale and market size, are not likely to be able to support much manufacturing and so access to wider African markets is essential. This implies some form of common market arrangement and a carefully planned distribution of industries within that organization so that international trade is encouraged. In turn this opens up a whole range of problems relating to common markets, economic co-operation, integration and eventually to the political dimension as well as the possible need for protection of infant industries. But if Africa is to develop out of its lowly status in the world economic order these problems will have to be faced up to.

Africa is ill served by its basic political geography which severely restricts development towards economic independence and prosperity. Much of that political geography derives from Africa's colonial past. Until its worst effects are permanently ameliorated, Africa will be condemned to poverty and dependency imposed by its colonial inheritance.

# REFERENCES

Ajibola, Judge (1994) *(Separate) Judgment, Territorial Dispute (Libya Jamahirya/ Chad,* The Hague: International Court of Justice.

Blake, Robert (1977) *A History of Rhodesia,* London: Methuen (extract from a speech by Jan Christiaan Smuts, August 1922).

Brownlie, Ian (1979) *African Boundaries: A Legal and Diplomatic Encyclopaedia,* London, Hurst (full text of an Exchange of Notes constituting an Agreement between the United Kingdom and Portugal regarding the Boundary between Tanganyika Territory and Mozambique, 11 May 1936; full text of an Organisation of African Unity Resolution adopted by the Assembly of Heads of State and Government, Cairo, 21 July 1964).

Du Bois, Felix (1897) *Timbuctoo the Mysterious,* London: William Heinemann.

Eybers, G. W. (1918) *Select Constitutional Documents Illustrating South African History 1795–1910,* London: George Routledge (full text of (Royal) Charter of the British South Africa Company, 29 October 1889).

Froude, J. A. (1886) *Oceana or England and her Colonies,* London: Longmans, Green, & Co.

Hall, R. N. and Neal, W. G. (1902) *The Ancient Ruins of Rhodesia (Monomotapae Imperium),* London: Methuen.

Hertslet, Sir E. (1909) *The Map of Africa by Treaty,* London: HMSO. 3rd edn of 1909, reprinted Frank Cass, 1967, (full text of the (Final) General Act of the Berlin Conference, 26 February 1885).

Hiller, V. W. (1949) 'The concession journey of Charles Dunell Rudd', in Constance E. Fripp and V. W. Hiller (eds), *Gold and the Gospel in Mashonaland 1888,* London: Chatto & Windus, Oppenheimer Series No. 4 (full text of The (Rudd) Concession signed by Lobengula, 30 October 1888).

Lugard, Frederick (1929) *The Dual Mandate in British Tropical Africa,* London.

McCullagh, Francis (1912) *Italy's War for a Desert Being some Experiences of a War Correspondent with the Italians in Tripoli,* London: Herbert & Daniel.

McEwen, A. C. (1971) *International Boundaries in East Africa,* Oxford: Clarendon Press.

Moffat, Robert (1842) *Missionary Labours and Scenes in Southern Africa,* London: John Snow.

Molyneaux, Henry Howard (Fourth Earl of Carnarvon) (1903) *Speeches on the Affairs of West Africa and South Africa,* London: John Murray, for private circulation.

Morel, E. D. (1915) *Ten Years of Secret Diplomacy: An Unheeded Warning,* London: The National Labour Press Ltd.

Nkrumah, Kwame (1963) *Africa Must Unite!,* London: William Heinemann.

196

Schapera, Isaac (ed.) (1961) *Livingstone's Missionary Correspondence 1841–1856,* London: Chatto & Windus (letter from David Livingstone to Arthur Tidman, Foreign Secretary, London Missionary Society, dated 17 March 1847).

UK (1952) *Basutoland, the Bechuanaland Protectorate and Swaziland: History of Discussions with the Union of South Africa 1909–1939,* London: HMSO, Cmd 8707.

Walton, Sir Edgar H. (1912) *The Inner History of the the National Convention of South Africa,* Cape Town: T. Maskew Miller (Sir George Grey's speech at the opening of the Cape Colony Parliament, 1859).

Wellington, J. H. (1967) *South West Africa and its Human Issues,* Oxford: Clarendon Press (Covenant of the League of Nations, Article 22, 1919; speech by D. Lloyd George, House of Commons, 3 July 1919).

# ANNOTATED
# BIBLIOGRAPHY

Asiwaju, A. I. (ed.) (1985) *Partitioned Africans: Ethnic Relations Across Africa's International Boundaries 1884–1984*, London; Hurst & Lagos, University of Lagos Press, xii, 275pp. Collection of twelve essays on the general theme of partitioned culture groups in Africa. A curate's egg of a book – good in parts.

Brownlie, Ian (1979) *African Boundaries: A Legal and Diplomatic Encyclopaedia*, London: Hurst, xxxvi, 1355pp. The 'bible' of African boundary studies. Each of over one hundred boundaries is dealt with systematically in authoritative detail.

Cervenka, Zdenek (ed.) (1973) *Land-locked Countries of Africa*, Uppsala, 369pp. Collection of papers mainly on the land-locked states of southern Africa. A little dated but there has been no replacement volume in this important field of interest.

Davidson, Basil (1992) *The Black Man's Burden: Africa and the Curse of the Nation State*, London: James Currey, xii, 355pp. paperback. Adopting a well-known radical title, this book is a stimulating but serious romp through Africa in the company of a long-experienced, well-informed, widely-travelled and much-respected Africanist whose own radicalism has not faded with the years.

Dumont, René (1963) *False Start in Africa*, London: Heinemann, 320pp. A classic chronicle of the misdirections of African development.

Grace, John and Laffin, John (1991) *Fontana Dictionary of Africa since 1960*, London: Fontana, xx, 395pp. A useful work for quick reference to post-independence African history.

Griffiths, Ieuan Ll. (1994) *The Atlas of African Affairs*, London and New York: Routledge, 2nd edn, xi, 233pp. paperback. The companion to the present volume, 'packed with forthright facts'. A wide-ranging introduction to African affairs with over 120 detailed maps and many tables of up-to-date, relevant statistics.

Hanlon, Joseph (1986) *Beggar your Neighbours: Apartheid Power in Southern Africa*, London: Catholic Institute for International Relations, James Currey & Indiana University Press, xii, 352pp., paperback. A detailed account of the apartheid state's military attacks on its neighbours and its stranglehold over them through its economic power and control of transport links.

Harden, Blaine (1992) *Africa: Dispatches From a Fragile Continent*, London: Fontana, 333pp., paperback. An American journalist's readable exposé of contemporary Africa based on personal experience. Something of a Dumont, if less authoritative.

Hargreaves, J. D. (1988) *Decolonization in Africa*, London: Longman, xvi, 263pp., paperback. A clear and concise account of a difficult and sometimes convoluted history which pays full attention to non-British as well as British practice in the decolonization process.

International Bank for Reconstruction and Development (World Bank) (annual), *World Bank Atlas*, Washington, DC: IBRD/WB. Readily available up-to-date source of key statistical data.

Keltie, John Scott (1893) *The Partition of Africa*, London: Stanford, xvi, 498pp. The best contemporary account of the European partition of Africa, giving special insights into the 'great game of scramble'.

McEwen, A. C. (1971) *International Boundaries of East Africa*, London: Oxford University Press, xii, 321pp. A detailed technical work on East African boundaries with meticulous attention to detail. Maps and full bibliography.

McLynn, Frank (1992) *Hearts of Darkness: The European Exploration of Africa*, London: Hutchinson, x, 390pp. A refreshing, wonderfully readable retelling of a well-known story full of new angles, interpretations and fascinating details.

Mandela, Nelson (1994) *Long Walk to Freedom: The Autobiography of Nelson Mandela*, London: Little, Brown & Co., 630pp. A personal chronicle of the long struggle against apartheid which vividly confronts the issues in straightforward, unequivocal terms.

Manning, Patrick (1988) *Francophone Sub-Saharan Africa 1880–1985*, Cambridge: Cambridge University Press, xii, 215pp., paperback. A reasonably up-to-date survey of francophone Africa that is both readable and informative with an extremely useful bibliographical essay.

Morel, E. D. (1920) *The Black Man's Burden*, Manchester: The National Labour Press, xii, 241pp. The wide-ranging culmination of a lifetime's campaign against the evils of European colonization in Africa by the dedicated and prolific author who exposed the gruesome horrors of the 'Congo system'.

Moyo, Sam, O'Keefe, Phil and Sill, Michael (eds) (1993) *The Southern African Environment: Profiles of the SADC Countries*, London: Earthscan, xiv, 354pp. Detailed profiles of the physical environment of the southern African states apart from South Africa. Not always consistent, with some chapters dated or patchy.

Mudenge, S. I. G. (1988) *A Political History of Munhumutapa, c1400–1902*, Harare, Zimbabwe Publishing House and London: James Currey, xxx, 420pp. A detailed history of pre-colonial Zimbabwe and its gold. With extensive bibiliography, statistical appendices and maps.

Omer–Cooper, John (1987) *History of Southern Africa*, London: James Currey, xiv, 298pp., paperback. Well illustrated, concise and clear account of the complex history of southern Africa sympathetically told. Full and useful bibliography.

Rogers, Barbara (1980) *Divide and Rule: South Africa's Bantustans*, London: International Defence and Aid Fund, revised edn, 136pp., paperback. The best brief, but authoritative, account of the enormity of 'Grand' apartheid, written before the monolith showed signs of disintegration.

Smith, Susanna (1990) *Front Line Africa: The Right to a Future*, Oxford: OXFAM, x, 387pp., paperback. The effect of the apartheid state's destablization of southern Africa through the 1980s. The poverty and suffering created in southern Africa by wars whipped up by the death throes of apartheid.

Vernet, Joel (ed.) (1994) *Pays du Sahel: Du Tchad au Senegal, du Mali au Niger*, Paris: Autrement, 231pp. An interesting collection (in French) of twenty-two essays by experts on the Franco-phone Sahelian countries, supported by statistical profiles and a bibliography of mainly French sources.

# INDEX

*Note*: The modern names of African places are mainly used, with cross-references from earlier names e.g. Zimbabwe (Southern Rhodesia). Sub-entries are generally in chronological order.